Alternative Energy Experiments

BY
SCHYRLET CAMERON AND CAROLYN CRAIG

ISBN 978-1-58037-520-7

Printing No. CD-404117

Visit us at www.carsondellosa.com

Mark Twain Media, Inc., Publishers
Distributed by Carson-Dellosa Publishing LLC

Table of Contents

Introduction to the Teacher

Alternative Energy Experiments explores the potential of using renewable resources as sustainable and long-term alternatives to conventional energy sources. As America seeks to curb its dependence on nonrenewable energy sources, such as oil and coal, science education will play a vital role in the transition to cleaner, more secure energy sources.

Alternative Energy Experiments is written for classroom teachers, parents, and students to explore the potential of renewable energy sources. Each section can be used as a full unit of study or as an individual lesson to supplement existing textbooks or curriculum programs. This book can be used as an enhancement to what is being done in the classroom or as a tutorial at home. The procedures and content background are clearly explained in an easy-to-read format that does not overwhelm the struggling reader. Vocabulary words are boldfaced, followed by a definition. **Mini Labs** can be completed individually or in a group setting. **Inquiry Labs** focus on the steps in the scientific method.

The easy-to-follow format of the book facilitates planning for the diverse learning styles and skill levels of middle-school students. National science standards addressed in each unit are identified and listed at the beginning of the book, simplifying lesson preparation. Each unit provides the teacher with alternative methods of instruction: reading exercises for concept development, simple hands-on activities to strengthen understanding of concepts, and challenging inquiry-based investigations provide opportunities for students to expand learning.

In our research for this book, several individuals, companies, and institutions of higher learning shared information concerning their work and research in the field of alternative energy. Sol Technologies, LLC, is a south Texas company specializing in solar light systems. They have partnered with several Texas middle schools in an effort to bring a hands-on solar energy program developed by Sol Technologies to students. Inventor Phil Sikes of Texas demonstrated many of the solar energy projects he is working on, including an easy-to-assemble, homemade solar-powered electric bicycle. Crowder College in southwest Missouri is recognized worldwide for its alternative energy program. Crowder's solar team designed and built the first solar-powered vehicle to successfully complete a coast-to-coast journey across the United States. Horse Hollow Wind Energy Center is a utilities company in Texas that uses wind turbines to produce electricity. It is the world's largest wind farm.

Alternative Energy Experiments supports the No Child Left Behind (NCLB) Act. The book promotes student knowledge and understanding of science and mathematics concepts through the use of good scientific techniques. The content, activities, and investigations are designed to strengthen scientific literacy skills that are correlated to the National Science Education Standards (NSES).

How to Use the Book

The format of *Alternative Energy Experiments* is specifically designed to facilitate the planning and teaching of science. Our goal is to provide teachers with strategies and suggestions to successfully implement each unit of study in the book.

Planning Guide

The Unit Planning Guide provides teachers with everything they need to organize and develop a lesson. It is intended to guide the teacher through the development and implementation of a lesson based on each student's abilities and needs. The Planning Guide includes:

1. National Standards: Each unit is correlated with the National Science Education Standards (NSES).

2. Skill Level: Each component of the unit is identified according to level of difficulty.
   - Level 1: concept development requiring basic reading skills
   - Level 2: hands-on activities requiring basic science skills
   - Level 3: inquiry-based activities requiring more advanced science skills

Challenging Units

The built-in flexibility of each unit accommodates a diversity of learning styles and skill levels. The format allows the teacher to begin the lesson with basic concepts and vocabulary presented in the reading exercise and expand to progressively more difficult hands-on activities and investigations. Each unit includes:

1. Student Information: introduces the concepts and essential vocabulary for the lesson in a directed-reading exercise.

2. Check Point: evaluates student comprehension of the information in the directed-reading exercise.

3. Mini Labs: strengthens student understanding of concepts with hands-on activities.

4. Inquiry Labs: explores concepts introduced in the directed-reading exercise through investigations focusing on steps in the scientific method.

Safety Tip: Adult supervision is recommended for all activities, especially those where chemicals, heat sources, electricity, or sharp or breakable objects are used. Safety goggles, gloves, hot pads, and other safety equipment should be used where appropriate.

Unit Planning Guide

National Science Standards Matrix

Each unit of study is designed to strengthen scientific literacy skills and is correlated with the National Science Education Standards (NSES).

Science Standards	Energy	Energy Sources	Electricity	Wind Energy	Solar Energy	Ocean Energy	Biomass Energy	Geothermal Energy	Hydro Energy	Nuclear Energy
	Units of Study									
Unifying Concepts	X	X	X	X	X	X	X	X	X	X
Content Standard A: Inquiry	X	X	X	X	X	X	X	X	X	X
Content Standard B: Physical Science	X	X	X	X	X	X	X	X	X	X
Content Standard C: Living Systems		X					X			
Content Standard D: Earth and Space	X	X		X	X	X		X	X	X
Content Standard E: Science and Technology	X	X	X	X	X	X	X	X	X	X
Content Standard F: Personal and Social Perspectives	X	X	X	X	X	X	X	X	X	X
Content Standard G: History and Nature of Science	X	X	X	X	X	X	X	X	X	X

Skill Level Matrix

Each unit contains four components designed to accommodate a diversity of learning styles and skill levels.

Unit Components	Level 1: Concept Development Basic Reading Skills	Level 2: Hands-on Activities Basic Science Skills	Level 3: Inquiry-Based Investigation Advanced Science Skills
Student Information	X		
Check Point	X		
Mini Labs		X	
Inquiry Labs			X

Inquiry Lab Scoring Guide

Category	4	3	2	1
Participation	Used time well, cooperative, shared responsibilities, and focused on the task.	Participated, stayed focused on task most of the time.	Participated, but did not appear very interested. Focus was lost on several occasions.	Participation was minimal OR student was unable to focus on the task.
Components of Investigation	All required elements of the investigation were correctly completed and turned in on time.	All required elements were completed and turned in on time.	One required element was missing/or not completed correctly.	The work was turned in late and/or several required elements were missing and/or completed incorrectly.
Procedure	Steps listed in the procedure were accurately followed.	Steps listed in the procedure were followed.	Steps in the procedure were followed with some difficulty.	Unable to follow the steps in the procedure without assistance.
Mechanics	Flawless spelling, punctuation, and capitalization.	Few errors.	Careless or distracting errors.	Many errors.

Comments:

Energy
Student Information

What Is Energy?

Energy is the capacity to do work or a source of usable power. All life needs energy to live and move. Most life on Earth gets its energy directly or indirectly from the sun. Vast amounts of energy are needed to support modern technological advancements and economic growth. A country's economic health depends on the availability of reliable and affordable energy.

> **Science by the Numbers**
>
> America has approximately 5% of the world's population but consumes about one-third of the world's energy.
>
> 40% of the world's electrical power is supplied by the solid fuel coal.

There are many forms of energy, but they can all be put into two groups: potential and kinetic. **Potential energy** is stored energy (chemical, nuclear, gravitational, electrical, and mechanical energy). **Kinetic energy** is moving energy (radiant, thermal, motion, and sound energy).

Fuel is any material that can be used to produce energy. For a fuel to be useful to humans, it must have two properties. 1) It can be stored until needed, and 2) when used, it can be controlled to produce work.

Early History of Energy Development

Fire produced the first source of energy for humans, and wood was the main fuel. Prehistoric humans discovered how to harness and use fire for cooking, heating, and protecting themselves from wild animals.

- In the 1800s, the primary fuel for homes and businesses was wood.
- In the late 1800s to mid 1900s, coal was considered the most important fuel.
- In the mid 1900s, electricity replaced wood heat in most homes and businesses.
- From the mid 1900s to today, oil has been the world's primary fuel source.

Uses of Energy

There are five main consumers of energy: homes, businesses, industries, transportation, and agriculture. The main uses of energy in homes and businesses are heating, lighting, and cooling. Industries use energy to generate steam and hot water for manufacturing. Gasoline and diesel fuel are used by almost 90% of all vehicles to transport people and goods. Different forms of energy (diesel fuel, gasoline, and propane) are used for different purposes in agriculture.

> **Just the Facts**
>
> OPEC (Organization of Petroleum Exporting Countries) was formed in 1960 by Iran, Iraq, Saudi Arabia, Kuwait, and Venezuela.
>
> Prior to the 1800s, people used oil-burning lamps for light. Whale oil (oil obtained from the blubber of various species of whales) was very popular because it burned with less odor and smoke than most other fuels.
>
> In the early 1970s, shortages of some fuels in the United States led to concerns that the nation might be facing an energy crisis. The result was a nationwide movement that advocated energy conservation and the development of alternate energy sources.

Kinds of Fuel

Fuels commonly used by humans can be divided into five main groups.

Solid Fuels

Coal, wood, peat, charcoal, and coke are the main sources of **solid fuel**. The use of some solid fuels, such as coal, is restricted or even prohibited in some cities because of the harmful gases released into the environment when they are burned.

Gas Fuels

Natural gas is the main **gas fuel** used in the United States. It comes from wells drilled deep underground and is piped to all parts of the United States. Propane and butane are gases produced from natural gas.

Liquid Fuels

Liquid fuel is made from crude oil, or petroleum. Crude oil comes from the ground. It is refined to create gasoline, kerosene, and other liquid fuels.

Atomic Fuels

Atomic fuels give off heat through fission or fusion of atoms. The fuel most widely used by nuclear plants for nuclear fission is uranium. Nuclear power plants produce electricity.

Chemical Fuels

Chemical fuels are very powerful man-made substances. These fuels are used to power jet engines and rockets.

Future Use of Energy

For the last century, the main source of energy has been **fossil fuels** (coal, oil, or natural gas) that are formed in the earth from the remains of ancient plants or animals. As the demand for fossil fuels has increased, three problems have occurred: increase in the cost of energy, exhaustion of fossil fuels, and damage to the environment. Scientists have begun exploring cleaner renewable alternatives to fossil fuel. Some of the types of energy sources that are being developed are solar, wind, hydroelectric, biomass, ocean, and geothermal energy.

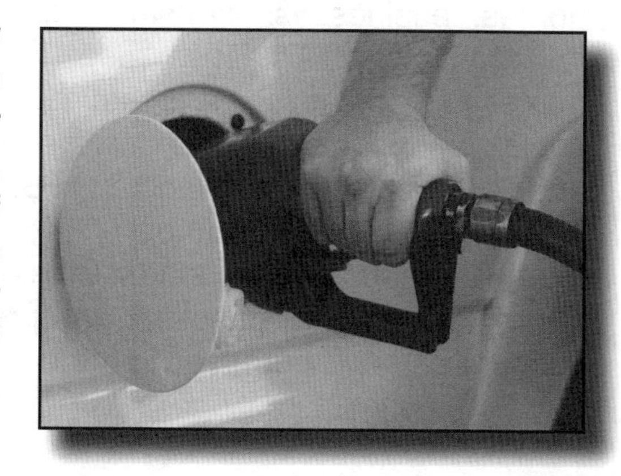

Name: ______________________ Date: ______________

Check Point

Matching

_____ 1. fossil fuels
_____ 2. fuel
_____ 3. energy
_____ 4. solid fuel
_____ 5. liquid fuel

a. coal, wood, peat, charcoal, and coke
b. a source of usable power
c. any material that can be used to produce energy
d. made from crude oil
e. formed in the earth from the remains of ancient plants or animals

Fill in the Blank

6. Atomic fuels give off heat through ______________ or ______________ of atoms.
7. Fire produced the first source of energy for humans, and ______________ was the main fuel.
8. Most life on Earth gets its energy directly or indirectly from the ______________.
9. ______________ ______________ is the main gas fuel used in the United States.
10. For the last century, the main source of energy has been ______________ ______________.

Multiple Choice

11. These fuels are used to power jet engines and rockets.
    a. atomic fuels
    b. chemical fuels
    c. liquid fuels
    d. solid fuels

12. This fuel supplies 40% of the world's electricity.
    a. wood
    b. natural gas
    c. coal
    d. crude oil

Constructed Response

Explain why a country's economic health depends on the availability of reliable and affordable energy. Give specific examples and details to support your answer. Answer on your own paper.

Name: ______________________ Date: ______________

Mini Labs

Mini Lab #1: Appliance Energy Use

Materials:
home appliances

Procedure: Knowing the amount of energy (watts) it takes to operate different appliances in your home can help you invest in more energy-efficient appliances. Choose five appliances found in your home. Record the appliance and wattage in the data table at right. You can usually find the wattage of most appliances stamped on the bottom or back of the appliance or on its nameplate.

Appliance	Wattage

Mini Lab #2: Energy Cost

Procedure: The Department of Energy reported that the average cost of residential electricity in 2009 was $0.12/kWh. You can calculate the amount of energy and the cost of energy used by different appliances in your home by using the formulas below. Complete the data table below. Calculate the cost for operating each appliance for two hours. Round to the nearest cent.

Step #1: How much energy does a 100-watt light bulb use in two hours?

Formula: $\frac{\text{POWER (watts) x TIME (hr)}}{\text{1,000 (watt x hr/kWh)}}$ *Example:* $\frac{\text{100 watts x 2 hours}}{\text{1,000 watt x hr/kWh}} = 0.2 \text{ kWh}$

Step #2: What is the cost of operating a 100-watt light bulb for 2 hours? Use $0.12/kWh as the average cost of electricity.

Formula: (kWh) x ($/kWh) *Example:* (0.2 kWh) x ($0.12/kWh) = $0.024

Appliance	Power (Watt)	Cost for Operating (2 hours)
1. Ceiling Fan	75	
2. Freezer	100	
3. Portable heater	1,500	
4. Television	250	
5. Clothes dryer	2,500	
6. Computer	360	
7. Water heater	3,200	

Name: ______________________________ Date: ______________________

Inquiry Lab: Saving Energy

Purpose: The purpose is a question that asks what you want to learn from the investigation. It should be clearly written, it usually starts with the verb "Does," and it can be answered by measuring something.

Purpose: *Does the method of bathing (shower or bath) affect the amount of energy (water) used?*

Research: The goal of the research is to find information that will help you make a prediction about what will occur in your experiment. Investigate water conservation and water usage of both baths and showers. Use the lines below for note taking.

____

____

____

____

____

____

____

____

____

____

____

____

Online Science: Learn more about water use in the United States at the following website. "Indoor Water Use in the United States." United States Environmental Protection Agency. <http://www.epa.gov/watersense/pubs/indoor.htm>

Hypothesis: Make an educated guess about what you think will happen in your project. Your hypothesis should be clearly written. It should answer the question stated in the purpose, be brief and to the point, and identify the independent and dependent variables.

Example: *It takes* (choose one) ***more*** *or* ***less*** *water to take a shower than a bath.*

Hypothesis: ____

____

Name: ______________________ Date: ______________

Procedure: The procedure is a plan for your experiment. The plan includes a list of the materials needed, step-by-step how-to directions (written like a recipe) for conducting the experiment, and it identifies the variables. Measurements are made and recorded using metric units.

Materials Needed:
bathtub with shower
meter ruler

Variables:
Independent: method of bathing
Dependent: amount of water used
Constants: same bathtub

Experiment:

Controlled Setup:

Step 1: Take a bath each day for one school week. Plug the drain. Run water in the tub to the level you would normally use for a bath.

Step 2: Place a meter ruler in the tub. Measure the amount of water in the tub from the bottom of the tub to the water level.

Step 3: Record the measurement in the data table below in centimeters. Drain the tub.

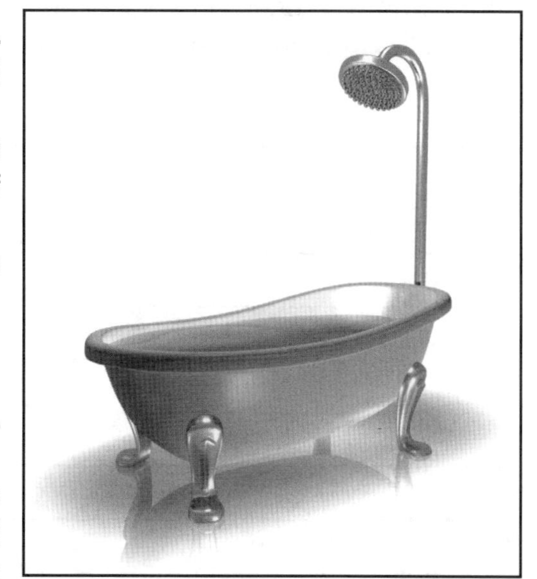

Experimental Setup:

Step 1: Take a shower every day for one school week. Plug the drain when taking the shower.

Step 2: Place a meter ruler in the tub. Measure the amount of water in the tub from the bottom of the tub to the water level when you finish showering.

Step 3: Record the measurement in the data table below in centimeters. Drain the tub

Results: Record the water level in the data table below.

Shower vs. Bath						
Controlled Setup (water used in cm)						
Method of Bathing	**Day 1**	**Day 2**	**Day 3**	**Day 4**	**Day 5**	**Average**
Bath						
Experimental Setup (water used in cm)						
Method of Bathing	**Day 1**	**Day 2**	**Day 3**	**Day 4**	**Day 5**	**Average**
Shower						

Name: ______________________________ Date: ____________________

Analysis: Study the results of your experiment. Create a graph that will compare the average water level in the control setup to the experimental setup. Place the dependent variable (water level) on the *y*-axis. Place the independent variable (bathing method) on the *x*-axis.

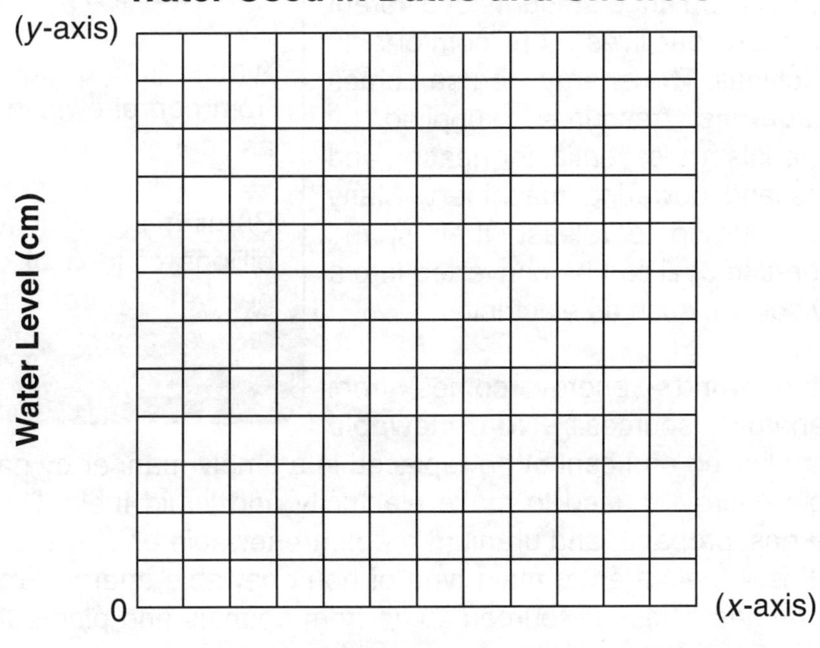

Method of Bathing

Conclusion: Write a summary of the experiment (what actually happened). It should include the purpose, a brief description of the procedure, and whether or not the hypothesis was supported by the data collected. Use key facts from your research to help explain the results. The conclusion should be written in first person ("I").

____

____

____

____

____

____

____

____

____

____

Energy Sources
Student Information

Energy Sources

Every day we use up huge amounts of different energy sources to make our lives more comfortable and to operate machines. The energy we use comes from **natural resources** (resources supplied by nature). These materials are essential for heating and lighting our homes and powering machinery. Many substances can be burned to release their stored energy. Energy stored in coal can be converted into a secondary energy source, such as electricity.

> **Science by the Numbers**
>
> Almost 90% of our energy comes from nonrenewable sources, such as coal.
>
> Burning of fossil fuels produces billions of tons of carbon dioxide per year. Carbon dioxide is one of the gases scientists believe contributes to global warming.

Most of the world's energy comes from nonrenewable energy sources. **Nonrenewable energy** sources are limited and cannot be replaced in a timely manner by natural processes. Most nonrenewable energy is used to make electricity and liquid fuels, like gasoline. Coal, petroleum, natural gas, propane, and uranium are nonrenewable energy sources. They come from the ground. Fossil fuels are the main type of nonrenewable energy. Fossil fuels include coal, oil, and natural gas. These resources come from animals and plants that died millions of years ago and then decomposed, creating a usable source of energy. Once these natural resources are used up, they are gone forever.

Renewable energy sources are not limited and can be replaced by natural processes. We use renewable energy sources mainly to make electricity. Renewable energy sources include biomass, geothermal, hydropower, solar, ocean, and wind energy.

History of Renewable and Nonrenewable Energy Use

Throughout history, humans have relied on a variety of energy sources, both renewable and nonrenewable, to make their lives more comfortable and to do work.

- Prehistoric humans burned wood for cooking, heating, and protecting themselves from wild animals.
- Many Native Americans used the energy from the sun to dry clothes and preserve foods, such as meat and fruits, to keep them from spoiling.
- From the early 1800s to the present, fossil fuels have become the main source of energy to heat and cool homes, power industries, and run transportation.
- In the early 1970s, shortages of some fuels in the United States led to a nationwide movement that advocated energy conservation and the development of alternative energy sources.
- In 2009, President Obama signed The American Recovery and Reinvestment Act, with provisions that called for building a clean, efficient energy supply based on renewable energy.

Just the Facts

In Hong Kong, the power used by one gym to operate the lights is produced by the people working out on such machines as the treadmill and stationary bicycle. The efforts of the patrons run a generator used to produce the electricity for the gym as they get physically fit.

Scientists are experimenting with the DNA of yeast. The new type of yeast excretes a substance similar to crude oil as a waste product. The oil can then be used as fuel.

Types of Energy Sources

Energy sources can be classified as renewable or nonrenewable.

Renewable Energy

1. Solar energy is produced by the sun. Nuclear reactions in the core of the sun give off radiant energy. Solar energy can be converted to thermal (heat) energy or electrical energy.
2. Biomass is plant or animal material or waste used to make electricity or fuel.
3. Geothermal energy harnesses the heat within the earth to heat buildings or generate electricity.
4. Hydro energy uses flowing water to spin turbines to generate electricity.
5. Wind energy is air in motion. It is used to spin large turbines to produce electricity.
6. Tidal/ocean energy is produced by the up and down motion of ocean waves.
7. Nuclear energy is energy trapped inside each atom. Scientists can split an atom apart, releasing tremendous energy. This is called nuclear fission. This energy can be used to generate electricity.

Nonrenewable Energy

1. Fossil fuels come from animals and plants that died millions of years ago and then decomposed, creating a usable source of energy.
2. Uranium is a metal found in some rocks. Nuclear power plants use uranium as fuel because its atoms are easily split apart.

Future Use of Energy

For the last century, the main source of energy has been **fossil fuels** (coal, oil, or natural gas) that are formed in the earth from the remains of ancient plants or animals. As the demand for fossil fuels has increased, three problems have occurred: an increase in the cost of energy, exhaustion of fossil fuels, and damage to the environment. Oil is the most common environmental pollutant. Scientists have begun exploring the use of renewable energy sources as a solution. These energy sources are often called **alternative energy resources**.

Advantages

- Renewable energy sources are inexhaustible resources that cannot be used up by humans.
- Renewable energy sources are environmentally friendly.

Disadvantages

- Nonrenewable energy sources are exhaustible resources and will be depleted (used up) in the near future.
- Nonrenewable energy sources, such as fossil fuels, produce pollutants when burned. This contributes to acid rain and global warming.

Name: ______________________________ Date: ______________

Check Point

Matching

_____	1. inexhaustible resource	a. limited and cannot be replaced in a timely manner by natural processes
_____	2. fossil fuels	b. unlimited resource
_____	3. nonrenewable energy	c. energy produced by the up and down motion of ocean waves
_____	4. tidal/ocean energy	d. resources supplied by nature
_____	5. natural resources	e. formed in the earth from the remains of ancient plants or animals

Fill in the Blank

6. For the last century, the main source of energy has been ______________________ ______________________.
7. Most nonrenewable energy is used to make ______________________ and ______________________ fuel.
8. Nonrenewable energy sources are ______________________ resources and will be depleted (used up) in the near future.
9. ______________________ energy is produced by the sun.
10. ______________________ ______________________ is one of the gases scientists believe contributes to global warming.

Multiple Choice

11. Which renewable energy source uses flowing water to create electricity?
    - a. solar energy
    - b. biomass energy
    - c. hydro energy
    - d. nuclear energy

12. Which is an example of a nonrenewable source of energy?
    - a. wind
    - b. flowing water
    - c. coal
    - d. ocean waves

Constructed Response

Explain the difference between inexhaustible resources and exhaustible resources. Give specific examples and details to support your answer. Answer on your own paper.

Name: ______________________ Date: ______________

Mini Labs

Mini Lab #1: Wind Speed

Materials:
compass
Beaufort Wind Scale (The Beaufort Wind Scale can be found at the following National Weather Service website <http://www.spc.noaa.gov/faq/tornado/beaufort.html>.)

Procedure: Record the wind speed and direction the same time each day for five days in the data table below. Write a description of the wind speed and the Beaufort Wind Scale indicator.

Date	Time	Direction	Description	Beaufort Scale

Conclusion: A wind turbine needs a minimum average wind speed of 13 mph to produce electricity economically throughout the year. Would a wind turbine work well in your area?

____

____

Mini Lab #2: Oil Spills

Materials:
large aluminum baking pan
rocks and gravel
motor oil
cotton balls
paper towels
yarn
water
plastic spoons
plastic drinking straws

Procedure: Make an ocean shoreline by placing rocks and gravel in one end of the pan. Add enough water to almost cover the rocks. Place several of the cotton balls on the rocks to represent birds and other animals. Add several drops of motor oil to the water to simulate an oil spill. Construct a boom to contain the oil. First, use yarn to try and contain the spill. Next, string plastic straws together and place them around the oil. Remove the boom and allow the oil spill to reach the shoreline. Try different methods to clean up the oil spill: paper towels, plastic spoons, and cotton balls.

Conclusion: Explain why oil spills have such a negative effect on a marine ecosystem.

____

____

____

Name: ______________________________ Date: ____________________

Inquiry Lab: Acid Rain and Plant Growth

Burning fossil fuels can emit pollutants into the air. This contributes to the formation of acid rain. This lab will help demonstrate how acid rain affects plants.

Purpose: The purpose is a question that asks what you want to learn from the investigation. It should be clearly written, it usually starts with the verb "Does," and it can be answered by measuring something.

Purpose: *Does the pH level of the water affect the root of a coleus plant?*

Research: The goal of the research is to find information that will help you make a prediction about what will occur in your experiment. Investigate the pH levels, composition, and formation of different types of soils. Also investigate how acid rain forms. Use the lines below for note taking.

____

____

____

____

____

____

____

____

____

____

____

Online Science: Learn more about acid rain at the following website. "Acid Rain." United States Environmental Protection Agency. <sun.http://www.epa.gov/acidrain/>

Hypothesis: Make an educated guess about what you think will happen in your project. Your hypothesis should be clearly written. It should answer the question stated in the purpose, be brief and to the point, and identify the independent and dependent variables.

Example: *The pH level of water* (choose one) ***will** or **will not** affect the root of a coleus plant.*

Hypothesis: ____

____

Name: ______________________________ Date: ______________

Procedure: The procedure is a plan for your experiment. The plan includes a list of the materials needed, step-by-step how-to directions (written like a recipe) for conducting the experiment, and it identifies the variables. Measurements are made and recorded using metric units.

Materials Needed:

2 small, clear jars
eye dropper
2 large containers
glass stirring rod
1 L distilled water
litmus paper and pH chart
15 mL white vinegar
1 g baking soda
beaker for measuring or graduated cylinder
2 cuttings of coleus (each with the same number of leaves and same length of stem)

Experiment:

Controlled Setup:

Step 1: Label one jar "neutral."

Step 2: Pour 500 mL of distilled water into one of the large containers. Use litmus paper and a litmus pH chart to measure the pH level of the neutral or control container. It should be 7.0. If it is higher, add a drop or two of vinegar, stir, and check it again. If it is lower than 7.0, sprinkle in a pinch of baking soda, stir, and check the pH again. Repeat until the color scale shows that the pH level is 7.0.

Variables

Independent: vinegar
Dependent: length of root
Constants: same type plant with same number of leaves and length of stem, water, sunlight, size of jars, same growing environment

Step 3: Fill the jar with the neutral water. Place a cutting in the jar. Make sure the stem and part of the lowest leaf are under water. Save any leftover water to keep the jar full.

Step 4: Place the jar in a sunny location.

Experimental Setup:

Step 1: Label one jar "acid."

Step 2: Pour 500 mL of distilled water into the other large container. Pour 15 mL of vinegar into the container, stir, and check the pH level. It should be 4.0. If it is higher or lower, add vinegar or baking soda as in Step 2 of the Controlled Setup until the pH level is 4.0.

Step 3: Fill the jar with the acid solution. Place the last cutting in the jar. Make sure the stem and part of the lowest leaf are under water. Save any leftover acid solution to keep the jar full.

Step 4: Place the jar in a sunny location.

Results: Every seven days, for two weeks, measure the root growth of the cuttings and record in the data table.

Controlled Group				Experimental Group			
	Day 7	Day 14	Average		Day 7	Day 14	Average
Root Length (cm)				Root Length (cm)			

Name: ______________________________ Date: ____________________

Analysis: Study the results of your experiment. Create a bar graph that will compare the average root length of the control group with the average root length of the experimental group. Place the dependent variable (root length) on the *y*-axis. Place the independent variable (pH level of watering solution) on the *x*-axis.

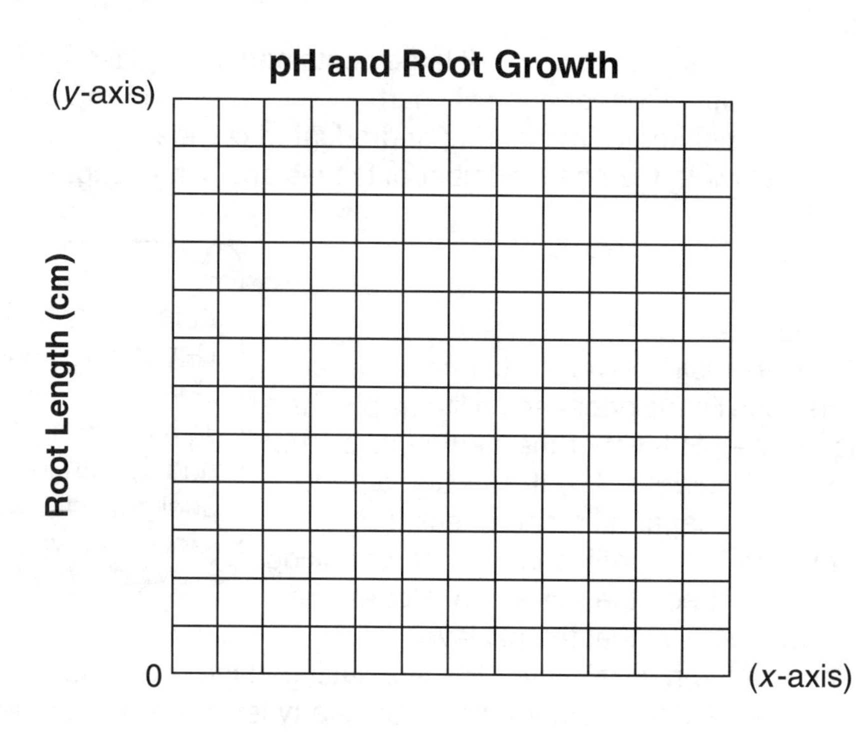

Conclusion: Write a summary of the experiment (what actually happened). It should include the purpose, a brief description of the procedure, and whether or not the hypothesis was supported by the data collected. Use key facts from your research to help explain the results. The conclusion should be written in first person ("I").

____

____

____

____

____

____

____

____

____

____

____

Electricity
Student Information

Electricity is the flow of electrical charges. Electricity is a secondary power source used to produce light and heat. It is also used to run motors. Electricity is a product of primary power sources, and it is neither renewable nor nonrenewable. We get electricity from the conversion of primary nonrenewable sources of energy, such as coal, natural gas, and oil, and primary renewable sources of energy, such as solar and wind.

Science by the Numbers

Approximately 25% of the world's population does not have access to electricity.

About 45% of a household's electricity consumption for one year is a result of trying to keep the home warm in the winter and cool in the summer.

Early History of Electricity

Until the late 1800s, people mostly used oil lamps and wood- or coal-burning stoves. Eventually, scientists discovered electricity. Their inventions using electricity have made our lives easier.

- William Gilbert (1544–1603) is credited with giving electricity its name.
- In 1752, Benjamin Franklin established a theoretical framework for the nature of electricity and electric charge.
- Dr. Luigi Galvani (1737–1798) of Italy made observations that led to the discovery of the electric battery.
- Michael Faraday (1791–1867) and Joseph Henry (1797–1878): Electric generators used today in power plants all over the world are based on the principles of induction outlined by Faraday and Henry.

How Electricity is Used

Electricity is relatively easy to make and send to where it is needed. It can be changed into other forms of energy, such as light and heat. Devices powered by electricity make our lives comfortable.

Electricity can be produced using turbines and generators. A primary source of energy powers a turbine to run a generator. The generator turns large copper coils inside huge magnets producing electricity. A transformer sends the electric current to the power lines. The electricity is carried to the users through wires to homes, businesses, and factories. The amount of electricity a power plant generates or a customer uses over a period of time is measured in **kilowatt-hours**.

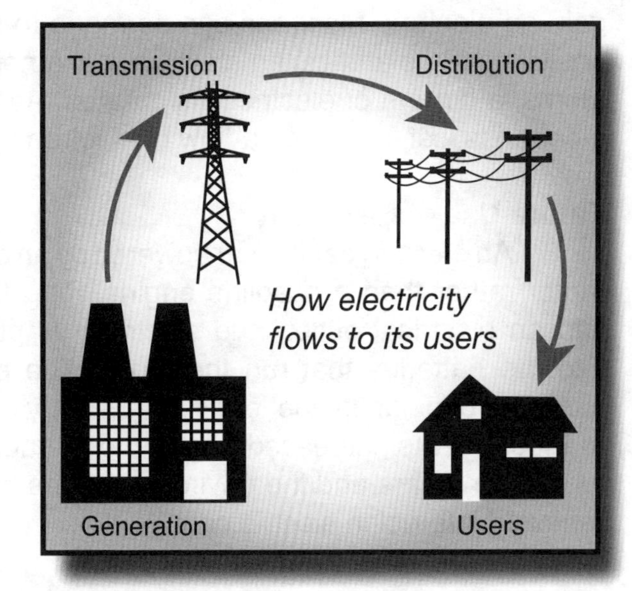

Just the Facts

In 1879, Thomas Edison invented the first long-lasting incandescent light bulb. Most people still use incandescent light bulbs to light their homes.

In 1891, President Benjamin Harrison had electricity installed in the White House. It is reported that he would not touch the light switches because he was afraid of getting shocked.

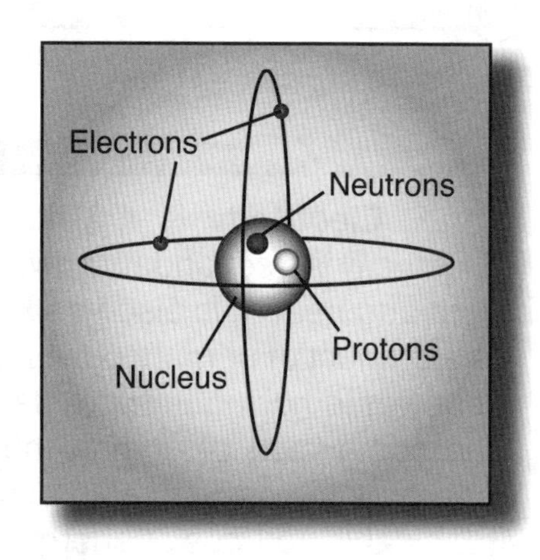

Every thing is made of matter. Matter is made up of tiny parts called **atoms**. Each atom has three even tinier parts. These parts are protons, electrons, and neutrons. The **nucleus** is the center of the atom. The **protons** and **neutrons** are small particles located in the nucleus of the atom. The protons have a positive (+) electrical charge. Neutrons are neutral. The neutrons have no charge. Protons and neutrons in an atom hold together very tightly. **Electrons** are small particles in orbit around the nucleus. The electrons have a negative (-) electrical charge. They are in orbit like the planets around the sun. The diagram above is the **Bohr Model** for atoms.

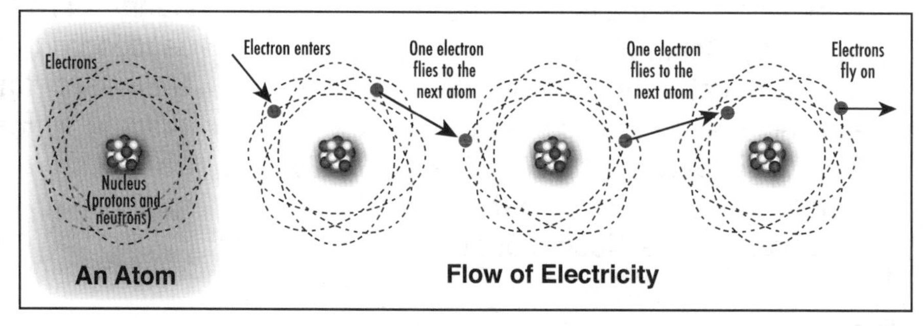

Electricity is made from the movement of electrons. The protons and electrons of an atom are attracted to each other. They both carry an **electrical charge**: protons have a positive charge (+) and electrons have a negative charge (-). Like magnets, opposite charges attract each other. When an atom is in balance, it has an equal number of protons and electrons. When atoms are not balanced, they need to gain an electron. Electrons can be made to move from one atom to another. (A proton with a positive charge attracts an electron with a negative charge). When an electron moves between atoms, a current of electricity is created. As one electron is attached to an atom and another electron is lost, it creates a flow of electrons.

Future Use of Electricity

An **electric car** is a car powered by an electric motor rather than a gasoline engine. Plug the car into an electrical outlet, and electricity is pumped into the batteries that run the motor. The electric car was popular in the early nineteenth century. In recent years, increased concern over fuel costs for gasoline cars and the environment has brought renewed interest in electric cars.

Name: ______________________ Date: ______________

Check Point

Matching

_____	1. electricity	a.	small particles in orbit around the nucleus
_____	2. protons	b.	credited with giving electricity its name
_____	3. William Gilbert	c.	the flow of electrical charges
_____	4. electrons	d.	units used to measure electricity used
_____	5. kilowatt-hours	e.	a positively charged particle located in the nucleus of an atom

Fill in the Blank

6. The ______________ turns large copper coils inside huge magnets, producing electricity.
7. The three parts of an atom are ______________, ______________, and ______________.
8. The protons and electrons carry an ______________ ______________.
9. In 1891, President ______________ ______________ had electricity installed in the White House.
10. The amount of electricity a power plant generates or a customer uses over a period of time is measured in ______________-______________.

Multiple Choice

11. An energy source powers a turbine to run a __________.
    a. motor b. transformer
    c. generator d. power plant

12. When a(an) __________ moves between atoms, a current of electricity is created.
    a. proton b. electron
    c. nucleus d. neutron

Constructed Response

Explain why electricity is considered a secondary power source. Give specific examples and details to support your answer.

____

____

____

____

Name: ______________________________ Date: ______________________

Mini Labs

Mini Lab #1: Meter Reading

Materials:

electric meter pencil paper an adult

Procedure: The electric meter at your house measures how many kilowatt-hours (kWh) of electricity your family uses. It may have several round dials or it may have a single display. Use the electric meter to calculate how much electrical energy your home uses in a day. Take a reading of your electric meter. Record the reading, along with the time, below in the data table. The next day, at the same time, take another reading. Record the reading, along with the time, below in the data table. To find the kilowatt-hours used in one day, subtract the meter reading on Day #1 from Day #2. Record the difference in the data table. **(Caution: Electricity is dangerous! Please have an adult help you take readings from your electric meter.)**

Day #1	Day #2	kWh Used in One Day

Conclusion:

Based on your results, determine how many kilowatt-hours of electricity your family would use in one month. ______________________

Mini Lab #2: Electrical Appliances

Materials:

pencil calculator

Procedure: A homeowner's electricity bill depends on the number of kilowatt-hours used each month. Below is a list of common household appliances. For the following activity, the appliances listed below are shown with the watt-hours that might be used. (The number of watts used by a given household appliance will vary from different manufacturers.) Calculate how long it will take each appliance to use one kilowatt-hour.

<u>Problem</u>: A coffeemaker uses 500 watts per hour. How long will it take for the coffeemaker to use one kilowatt-hour? Divide 1,000 (1,000 watts equals one kilowatt-hour) by 500 (watts used per hour by the coffeemaker).

Example: 1,000/500 = 2 hours

Appliance	Watt-hours	Time
1. Toaster	200 watts per hour	
2. Clothes Dryer	2,000 watts per hour	
3. Television	400 watts per hour	
4. Refrigerator	250 watts per hour	

Name: ______________________________ Date: ____________________

Inquiry Lab: Light Bulbs

Purpose: The purpose is a question that asks what you want to learn from the investigation. It should be clearly written, it usually starts with the verb "Does," and it can be answered by measuring something.

Purpose: *Does the type of light bulb used in a lamp affect the amount of heat energy released?*

Research: The goal of the research is to find information that will help you make a prediction about what will occur in your experiment. Investigate indoor lighting, incandescent bulbs, fluorescent bulbs, and LED bulbs. Use the lines below for note taking.

Online Science: Learn more about light bulbs at the following website. "Lighting Products." U.S. Department of Energy. <http://www.energystar.gov/index.cfm?c=lighting.pr_lighting>

Hypothesis: Make an educated guess about what you think will happen in your project. Your hypothesis should be clearly written. It should answer the question stated in the purpose, be brief and to the point, and identify the independent and dependent variables.

Example: *The type of light bulb used* (choose one) ***will*** *or* ***will not*** *affect the amount of heat energy released.*

Hypothesis: ____

Name: ______________________ Date: ______________

Procedure: The procedure is a plan for your experiment. The plan includes a list of the materials needed, step-by-step how-to directions (written like a recipe) for conducting the experiment, and it identifies the variables. Measurements are made and recorded using metric units.

(Caution: Fluorescent bulbs contain a small amount of mercury. Adult supervision is recommended for this experiment.)

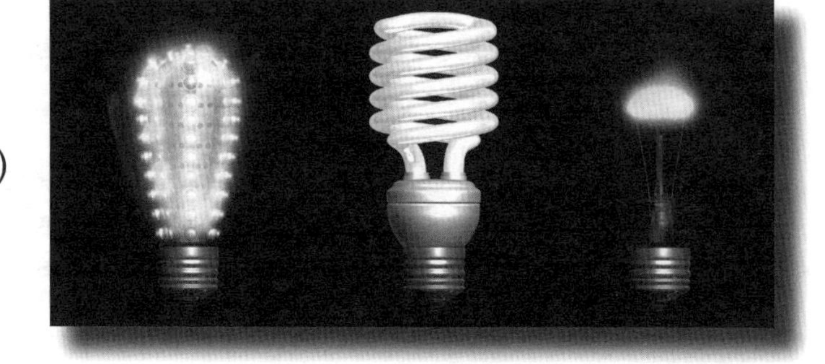

Materials Needed:
incandescent bulb (60-watt)
compact fluorescent bulb (equal to 60 watts)
LED bulb (equal to 60 watts)
thermometer
lamp

Experiment:

Controlled Setup:

Step 1: Place an incandescent light bulb in the lamp.
Step 2: Turn the incandescent light bulb on.
Step 3: Hold a thermometer 15 cm above the bulb for one minute. Record the temperature in the data table below. Turn the bulb off and let it cool.
Step 4. Repeat steps 2 and 3 two more times.

Variables
Independent: type of bulb
Dependent: heat energy
Constants: same lamp, thermometer, voltage, length of time measuring temperature

Experimental Setup:

Step 1: Place a fluorescent light bulb in the lamp.
Step 2: Turn the fluorescent light bulb on.
Step 3: Hold a thermometer 15 cm above the bulb for one minute. Record the temperature in the data table below. Turn the bulb off and let it cool.
Step 4. Repeat steps 2 and 3 two more times.
Step 5: Repeat steps 1 through 4 using the LED bulb.

Results: Record the temperatures (°C) in the data table below. Average the trials for each type of bulb.

Control Set Up				
Type of Bulb	**Trial #1**	**Trial #2**	**Trial #3**	**Average**
Incandescent				
Experimental Setup				
Type of Bulb	**Trial #1**	**Trial #2**	**Trial #3**	**Average**
Compact Fluorescent				
LED				

Name: ______________________________ Date: ______________

Analysis: Study the results of your experiment. Create a graph that will compare the average temperature in the control group with the experimental group. Place the dependent variable (temperature) on the *y*-axis. Place the independent variable (type of bulb) on the *x*-axis.

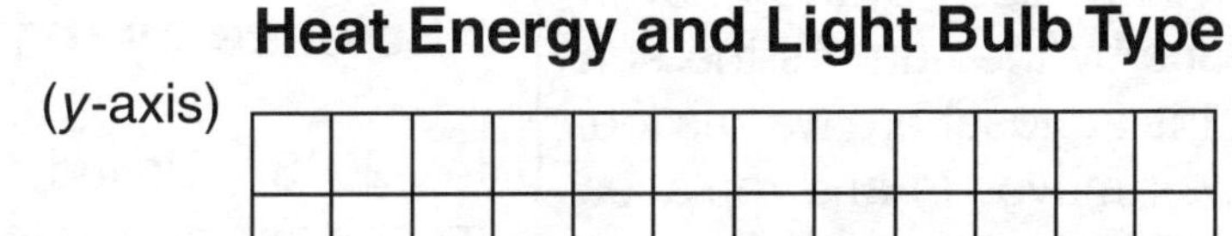

Heat Energy and Light Bulb Type

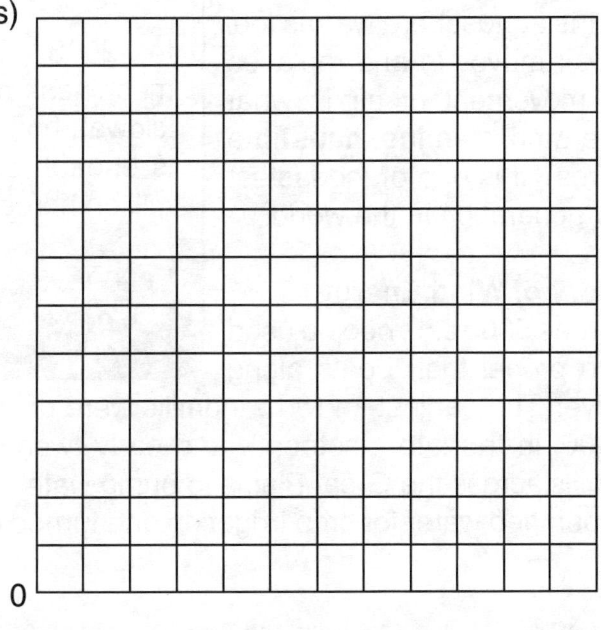

Conclusion: Write a summary of the experiment (what actually happened). It should include the purpose, a brief description of the procedure, and whether or not the hypothesis was supported by the data collected. Use key facts from your research to help explain the results. The conclusion should be written in first person ("I").

Wind Energy
Student Information

Wind, air in motion, is a result of the uneven heating of the earth's surface by the sun. As the sun warms the earth's surface, the atmosphere warms too. Warm air rises. Then cool air moves in and replaces the rising warm air. This movement of air is what makes the wind blow. Since wind is an **inexhaustible resource** (unlimited resource), it is one of the fastest growing forms of electricity generation in the world.

Science by the Numbers

In 2008, United States wind machines generated 52 billion kilowatt-hours of electricity. This is enough electricity to serve 4.6 million homes.

In 2008, 31 U.S. states used wind turbines to generate electricity.

Early History of Wind Energy

As early as 5000 B.C., people used the wind to propel their boats along the Nile River. The earliest-known windmills were developed in Persia around A.D. 500–900. In the late nineteenth and early twentieth centuries, pioneers built windmills across the Great Plains to pump water from underground wells. Windmills pumped water for crop irrigation and turned millstones to grind grain into flour.

- In 1854, Daniel Halladay designed a windmill with an open tower and thin wooden blades.
- In 1883, Thomas O. Perry conducted over 5,000 tests on 61 different models of wind wheel before developing the Aermotor, a windmill made entirely of steel. The invention was nicknamed the mathematical windmill because of all the study and calculation that Perry did.
- In 1888, Charles F. Brush designed the first windmill that generated electricity.
- In 1919, wind energy theory was developed by German physicist Albert Betz and published in his book *Wind Energy and its Use in Windmills.*

How Wind Energy is Used

Kinetic energy (moving energy) can be collected from the wind and converted to electrical energy.

Today, wind turbines are being built to provide electrical energy to power homes and businesses. Wind produces kinetic energy that turns the blades of a turbine; this drives a generator that produces electricity. Electricity travels through a transformer and into local electrical networks through transmission lines, which distribute electricity to homes.

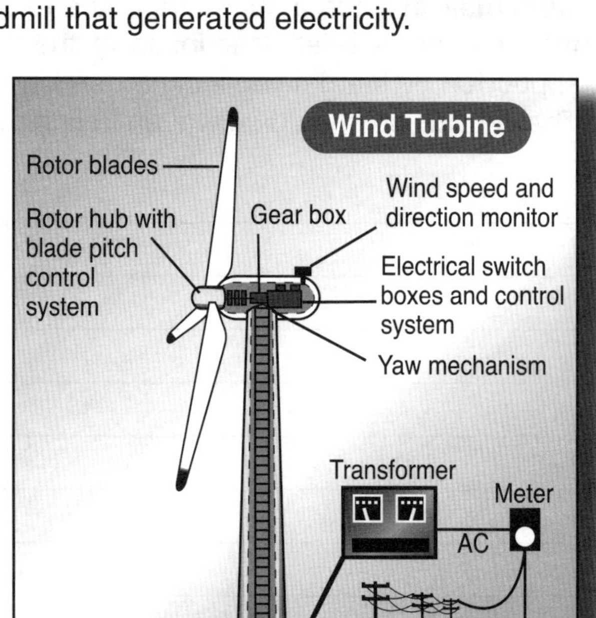

Just the Facts

Americans used small windmills until the late 1920s to generate electricity in rural areas without electric service.

The Horse Hollow Wind Energy Center in Texas is the world's largest wind farm. The 421 wind turbines generate enough electricity to power 220,000 homes per year.

According to the Department of Energy, offshore wind farms could provide enough energy to power the entire nation.

Types of Wind Turbines

There are two types of wind turbines: the horizontal-axis and the vertical-axis. The **horizontal-axis wind turbines** have either two or three blades. A typical horizontal wind turbine is as tall as a 20-story building and has three blades that span 200 feet across. They operate with the blades facing the wind. The **vertical-axis wind turbines** look like a giant, two-bladed egg beater. They have blades that go from top to bottom, and are 100 feet tall and 50 feet wide. Most wind machines being used today are the horizontal-axis type.

Placement of Wind Turbines

Wind turbines must be carefully placed in order to capture as much wind as possible. Turbines operate best in areas where wind speeds are 16 mph to 20 mph at a height of 50 meters. Good site locations are the tops of rounded hills, wide-open plains, shorelines, and mountain gaps that produce wind funneling. Ideal placement of wind turbines takes advantage of **prevailing winds** (winds that dependably blow from a certain direction). The United States has offshore and coastal areas that are suitable for wind development. Wind turbines can be sited offshore where the water is shallow and winds blow harder. This means that turbines built further offshore should capture more wind energy. A major wind farm has been proposed off the coast of Cape Cod, one of the largest suitable offshore areas in the United States.

Future Use of Wind Energy

In the near future, scientists believe wind energy will be the most cost-effective source of electrical power. Advances in wind turbine design will enable turbines to operate at lower wind speeds and to harness more of the wind's energy. It is estimated that wind power alone could be used to meet all of the energy needs of the United States.

Advantages

- Wind energy is clean, renewable, and doesn't produce greenhouse gases.
- Wind energy is free.

Disadvantages

- Wind farms require large, open spaces with huge windmills that some people think are unattractive.
- Windmills could interfere with the flight paths of migratory birds.

Name: ______________________ Date: ______________

Check Point

Matching

_____ 1. inexhaustible resource		a. air in motion
_____ 2. wind		b. unlimited resource
_____ 3. kinetic energy		c. have either two or three blades
_____ 4. vertical-axis wind turbine		d. looks like a giant, two-bladed egg beater
_____ 5. horizontal-axis wind turbine		e. moving energy

Fill in the Blank

6. ______________ produces kinetic energy that turns the blades of a turbine; this drives a ______________ that produces electricity.
7. The earliest-known windmills were developed in ______________ around A.D. 500–900.
8. In 1888, Charles F. Brush designed the first windmill that generated ______________.
9. Wind energy is clean and renewable and doesn't produce ______________ gases.
10. It is estimated that wind power can meet all of the United States' ______________ needs.

Multiple Choice

11. In which state is the world's largest wind farm located?
    a. California b. Texas
    c. Arizona d. New York

12. Who invented the mathematical windmill?
    a. Thomas O. Perry b. Charles F. Brush
    c. Daniel Halladay d. Charles Fritts

Constructed Response

Explain why scientists consider the wind an inexhaustible resource. Give specific examples and details to support your answer.

___

___

___

___

___

___

Name: ______________________ Date: ______________

Mini Labs

Mini Lab #1: Pinwheel

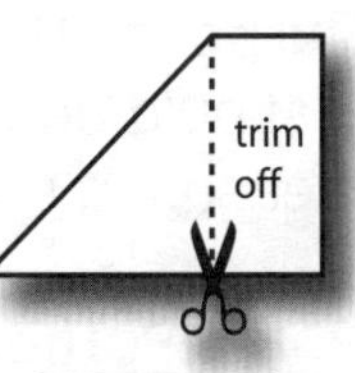

Materials:

scissors push pin fan
unsharpened pencil with an eraser construction paper

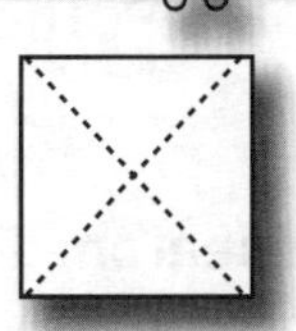

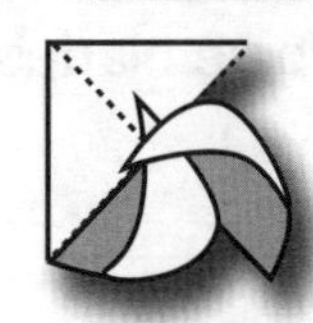

Procedure: Make the paper into a square. Take one edge of the paper and fold it to meet the top. You will have a large triangle and a small rectangular strip. Cut off the rectangular strip. Once you have your square, fold it corner to corner, unfold it and fold it corner to corner on the opposite side, forming a folded "X". Using your scissors, cut along the fold lines, stopping 3⁄4 of the way down the fold. Carefully gather every other corner and bring it down to the center. Try not to fold it when bringing it to the center. Folds will interfere with the pinwheel's ability to spin. Insert the push pin into the center, making sure you go through all corners. Without letting go of the push pin and paper, push the pin through the side of the eraser. Place your pinwheel in front of a running fan. Notice what happens when you turn your pinwheel in different directions.

Conclusion: Why would the placement of a wind turbine be important?

____

____

Mini Lab #2: Wind Sock

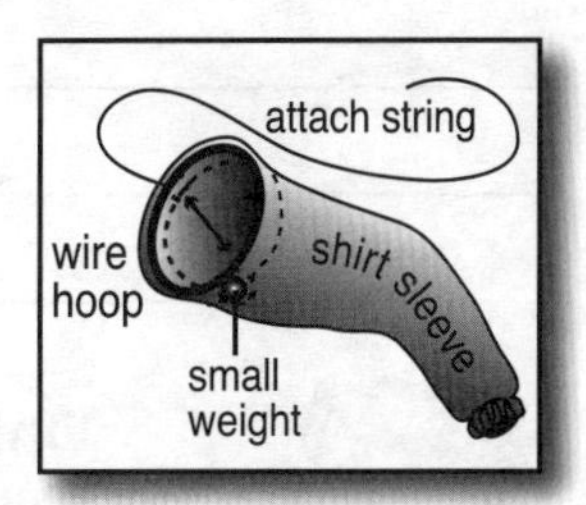

Materials:

shirt sleeve (old windbreaker works great) scissors wire snips
a weight 1 meter string
100 cm lightweight wire tape measure needle and thread

Procedure: Measure the outer part (circumference) of the large opening of the sleeve. Measure and cut the wire the same length. Form a circle with the wire. Fold the opening of the sleeve over the wire, forming a 2-cm band. Stitch around the sleeve, enclosing the wire in the band, leaving a small opening for the weight. Place the weight in the band. Secure the weight in place by stitching around it. Sew the string to the band opposite the weight. Tie your wind sock outside to a flag pole. The weight you sewed in the material should hold the mouth of your wind sock facing into the wind. Observe your wind sock over several days. Use a compass and the wind sock to determine the direction of the wind each day.

Conclusion: Why would the placement of a wind turbine be important?

____

____

Name: ______________________________ Date: ____________________

Inquiry Lab: Wind Turbine

Purpose: The purpose is a question that asks what you want to learn from the investigation. It should be clearly written, it usually starts with the verb "Does," and it can be answered by measuring something.

Purpose: *Does the number of blades affect the amount of voltage produced by a wind turbine?*

Research: The goal of the research is to find information that will help you make a prediction about what will occur in your experiment. Investigate wind energy, wind turbines, and types of turbine blades. Use the lines below for note taking.

____

____

____

____

____

____

____

____

____

____

____

____

Online Science: Learn more about wind energy at the following website. "Wind Basics—Energy from Moving Air." U.S. Department of Energy. <http://www.eia.doe.gov/kids/energyfacts/sources/renewable/wind.html>

Hypothesis: Make an educated guess about what you think will happen in your project. Your hypothesis should be clearly written. It should answer the question stated in the purpose, be brief and to the point, and identify the independent and dependent variables.

Example: *The number of blades* (chose one) ***will*** *or* ***will not*** *affect the amount of voltage produced by a wind turbine.*

Hypothesis: ____

____

Name: ______________________________ Date: ______________

Procedure: The procedure is a plan for your experiment. The plan includes a list of the materials needed, step-by-step how-to directions (written like a recipe) for conducting the experiment, and it identifies the variables. Measurements are made and recorded using metric units.

Materials Needed:

1 small motor	scissors	glue
4 regular paper clips	metric ruler	wire cutters
1 rubber band	voltmeter	cardboard
2 alligator clips	cork	fan

2 60-cm pieces of electrical wire with 2 cm of insulation removed from the ends

Variables
Independent: number of blades
Dependent: voltage produced
Constants: same motor, rubber band, alligator clips, paper clips, voltmeter, cork, cardboard, glue, tape, ruler, and wiring

Experiment: (Caution: Wear safety goggles.)

Controlled Setup:

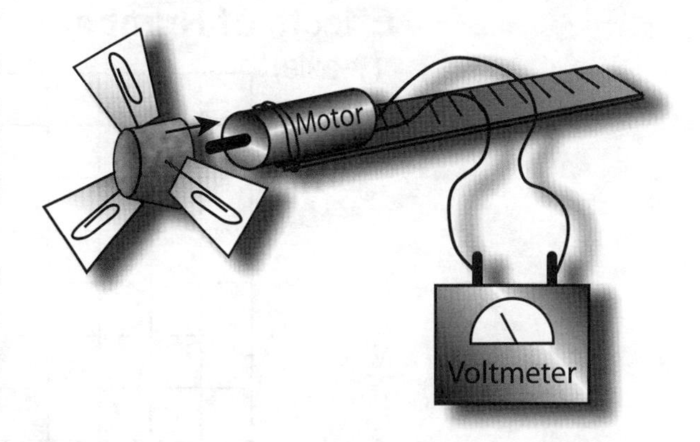

Step 1: Attach the motor to one end of the ruler by wrapping it with a rubber band. Leave the motor shaft extending beyond the end of the ruler.

Step 2: Tape the wires to the ruler at the end without the motor.

Step 3: Attach the other two ends of the wire to the alligator clips.

Step 4: Straighten out a large paper clip, leaving one end with the hook. Cut all 4 clips the same length.

Step 5: Cut 4 pieces of cardboard into 2-cm X 5-cm strips.

Step 6: Glue the hooked part of a paper clip to the bottom of each cardboard piece. Leave 1 cm of the clip sticking out beyond the cardboard. (Allow glue to dry completely).

Step 7: Using the paper clip, insert 2 blades about 5 mm from the small end of the cork. Equally space the blades around the circumference of the cork. Use a protractor to set the blades at a 45 degree angle to the ruler.

Step 8. Make a hole smaller than the motor shaft in the center of the large end of the cork. Carefully push the cork onto the motor's shaft.

Step 9: Connect the alligator clips to the voltmeter. Place your turbine 30 cm from a fan. Turn on the fan. Record the voltage produced in the data table on the next page. Repeat the experiment 2 more times and record the voltage. Calculate the average voltage produced and record the value in the data table.

Experimental Setup:

Repeat Steps 7–9 with 3 blades and then with 4 blades.

Results: In the chart on the next page, record the voltage in amps for each test and calculate the average amps produced by the blades.

Name: ______________________ Date: ______________

Control Group Voltage				
Blades	**Test #1**	**Test #2**	**Test #3**	**Average**
2 Blades				
Experimental Group Voltage				
Blades	**Test #1**	**Test #2**	**Test #3**	**Average**
3 Blades				
4 Blades				

Analysis: Study the results of your experiment. Create a graph that will compare the voltage produced with the number of blades used. Place the dependent variable (voltage produced) on the *y*-axis. Place the independent variable (number of blades) on the *x*-axis.

Effects of Number of Blades on Voltage Output

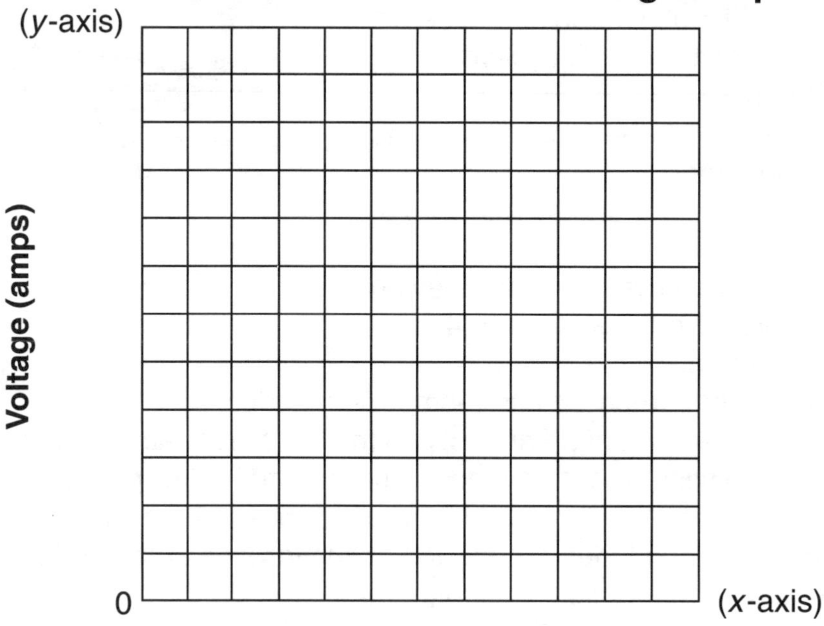

Conclusion: Write a summary of the experiment (what actually happened). It should include the purpose, a brief description of the procedure, and whether or not the hypothesis was supported by the data collected. Use key facts from your research to help explain the results. The conclusion should be written in first person ("I").

Solar Energy
Student Information

Solar Radiation

The sun is a big ball of gas that produces energy resulting from nuclear reactions in its core. In the core, hydrogen atoms combine to form helium atoms. This gives off radiant energy. The radiant energy travels to Earth as heat and light. Because scientists predict the sun is capable of producing energy for billions of years, the sun is considered an **inexhaustible resource** (unlimited resource). Energy from the sun's rays, called solar radiation, that reaches Earth is called **solar energy**. The two major types of solar energy are heat and light.

Science by the Numbers

The sun produces more energy in 60 minutes that the entire population of the world can use in one year.

The sun is 93 million miles away from Earth. It takes sunlight 8 minutes to travel the distance to Earth.

Early History of Solar Energy

Ancient people built their homes to face the warm winter sun, and they used the sun to dry clothes and to preserve foods, such as meat and fruits, to keep them from spoiling. Eventually, scientists discovered ways to harness energy from the sun. Their inventions laid the groundwork for later development of solar technology.

- In 1767, Horace-Bénédict de Saussure made the first solar oven, which he called a "hot box."
- In 1772, Antoine-Laurent Lavoisier invented a solar furnace to melt metal. It even melted a diamond!
- In 1882, Augustin Mouchot and Albe Pifre invented a solar printing press. It printed 500 pages an hour.
- In 1883, Charles Fritts described the first solar cell made from selenium.
- In 1830, Sir John Herschel used a solar cooker to cook food during his South African expedition.
- In 1908, William J. Bailey made a solar collector with copper coils.

How Solar Energy is Used

Solar energy can be converted to **thermal** (heat) energy or electrical energy.

Thermal Energy: **Solar panels** are devices that can be placed on buildings to absorb and collect solar radiation to heat water. The panel is a glass-covered box. The box is painted black inside. A black pipe runs through the box. The black pipe absorbs the sunlight, and the water flowing through the pipe is heated.

Electrical Energy: Sunlight is collected by a **photovoltaic** (PV) device or solar cell and is transformed into a small amount of electrical energy. Solar cells are used to power watches, decorative landscape lighting, calculators, and satellites. *Sojourner,* a solar-powered, remote-controlled robot, used energy transformed by a panel of solar cells to explore Mars.

Just the Facts

The world's largest solar power plant is located in the Mojave Desert in California and Nevada. It covers 1,000 acres with solar reflectors.

Albert Einstein won the Nobel Prize in 1921 for his experiments with photoelectric effects.

Using solar cells, the Sahara Desert could generate all the world's electricity requirements.

All TV and communications satellites are powered by solar energy using photovoltaic cells.

President Jimmy Carter installed solar panels on the White House in 1979.

Collecting Solar Energy

Solar Dish: A mirrored solar dish that looks like a home satellite receiver uses concentrated sunlight to produce electricity.

Solar Ponds: A pond of saltwater collects and stores solar thermal energy. Sunlight heats the salty water, and it sinks to the bottom of the pond. The hot water can be used to supply homes with hot water or to make electricity.

Solar Farms: Thousands of panels of solar cells are placed in rows. The panels are tilted toward the sun to collect thermal energy that is then transformed into electricity. It is carried to homes through overhead power lines.

Solar Power Plants: Rows of mirrors collect the sun's energy. The energy is used to heat water to make steam. The steam is used to power **generators**, machines that produce electricity. California and Arizona both have solar power plants that generate electricity.

Solar Power Tower: Thousands of mirrors reflect heat to the tower, which is filled with liquid. The heat is used to make steam. The steam is used to power generators that make electricity.

Future Use of Solar Energy

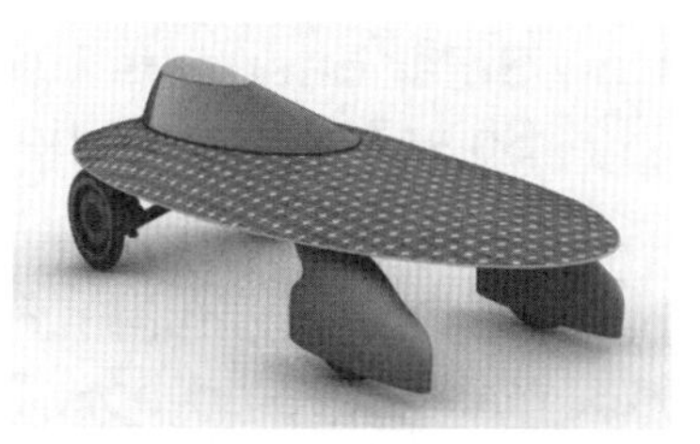

Solar-powered transportation is being developed as an alternative to fossil-fueled transportation. Centurion is an experimental solar-powered aircraft developed by NASA. Engineers and scientists are looking at ways to make solar power an efficient and economical way to fuel cars, bicycles, and motorcycles.

Advantages

- Solar energy is free, inexhaustible, and clean.
- The use of solar energy reduces our dependence on foreign sources of energy.

Disadvantages

- Solar energy is unreliable. The amount of sunlight that arrives on the earth's surface is affected by many factors, including weather, air pollution, time of day, and time of year.
- Solar technology can be expensive.

Name: ______________________________ Date: ____________________

Check Point

Matching

_____ 1. inexhaustible resource
_____ 2. solar energy
_____ 3. thermal energy
_____ 4. photovoltaic
_____ 5. generator

a. heat energy
b. unlimited resource
c. machine used to produce electricity
d. solar cell
e. solar radiation

Fill in the Blank

6. Solar energy can be converted to ______________ energy or ______________ energy.
7. A solar ______________ looks like a home satellite receiver.
8. ______________ ______________ use salt water to collect and store solar thermal energy.
9. Solar-powered transportation is being developed as an alternative to ______________ -______________ transportation.
10. The use of solar energy reduces our dependence on ______________ sources of energy.

Multiple Choice

11. Which president added solar panels to the White House?
    a. Ronald Reagan
    b. Barack Obama
    c. Jimmy Carter
    d. George W. Bush
12. Who made the first solar oven?
    a. Sir John Herschel
    b. Horace-Bénédict de Saussure
    c. Antoine-Laurent Lavoisier
    d. Charles Fritts

Constructed Response

Explain why scientists consider the sun a renewable resource. Give specific examples and details to support your answer.

____

____

____

____

Name: ______________________ Date: ______________

Mini Labs

Mini Lab #1: Solar Cooker

Materials

pizza or shoe box
aluminium foil
plastic wrap
scissors
black spray paint
marshmallows
graham crackers
chocolate candy bar
black construction paper
ruler

Procedure: Spray paint the outside of the box black. After the paint is dry, draw an 8 inch x 11 inch square on the lid of the box. Cut out three sides of the square, and fold the flap back along the uncut edge to make a window in the cooker. Glue aluminum foil to the inside of this flap. Cover the window with a piece of plastic wrap. Tape the wrap to the box lid. Glue black construction paper to the inside bottom of the box. Glue foil to the inside walls of the box. Place a graham cracker in the box; stack two chocolate squares and a marshmallow on top of the cracker. Prop the box at an angle facing the sun. Use a ruler to prop the flap open. When the chocolate has melted (10 to 15 minutes), remove the s'more, place a cracker on top of the marshmallow, and enjoy.

Conclusion: Why is black paper and paint used to construct the solar cooker? ______________

____

____

Mini Lab #2: Solar Collector

Materials

6 plastic bottles
6 different colors of spray paint, including black and white
6 seven-inch balloons

Procedure: Spray paint one bottle black and the other white. Spray paint each of the remaining bottles a different color. Place the open end of one balloon over the mouth of each bottle. Make sure the balloon forms an airtight seal. Place each bottle in direct sunlight. Observe the bottles for 30 minutes. Record your observations.

Observations:

1. What do you notice happening to the balloons? ______________

   ____

2. Touch each bottle. What do you notice? ______________

   ____

Conclusion: ____

____

Name: ______________________________ Date: ______________________

Inquiry Lab: Solar Water Heater

Purpose: The purpose is a question that asks what you want to learn from the investigation. It should be clearly written, it usually starts with the verb "Does," and it can be answered by measuring something.

Purpose: *Does the color of the tubing affect the temperature of the water in a solar collector?*

Research: The goal of research is to find information that will help you make a prediction about what will occur in your experiment. Investigate solar energy, solar hot water heaters, and solar collectors. Use the space below for taking notes.

____

____

____

____

____

____

____

____

____

____

____

Online Science: Learn more about solar energy at the following website.
"Solar Basics—Energy from the Sun." U.S. Department of Energy.
<http://www.eia.doe.gov/kids/energyfacts/sources/renewable/solar.html>

Hypothesis: Make an educated guess about what you think will happen in your project. Your hypothesis should be clearly written. It should answer the question stated in the purpose, be brief and to the point, and identify the independent and dependent variables.

Example: *The color of the tubing* (choose one) ***will*** *or* ***will not*** *affect the temperature of the water in the solar collector.*

Hypothesis: ____

____

____

Name: ______________________ Date: ______________

Procedure: The procedure is a plan for your experiment. The plan includes a list of the materials and step-by-step how-to directions (written like a recipe) for conducting the experiment, and it identifies the variables. Measurements are made and recorded using metric units.

Materials Needed:

1 clothespin
plastic wrap
Celsius thermometer
1 shoe box
metric ruler
3 liters of tap water
3 meters of each color of plastic tubing: black, clear, and white
two 1-liter plastic containers

Experiment

Controlled Setup:

Step 1: Punch one hole big enough for the tubing to pass through in each end of the shoe box, and fill the box with crumpled plastic wrap.
Step 2: Insert the clear tubing through the box with 100 cm left sticking out from each end.
Step 3: Place the remaining 100 cm of tubing on top of the plastic wrap in a zigzag pattern from one side to the other of the box.
Step 4: Cover the box with clear plastic wrap.
Step 5: Place the box in direct sunlight with one end elevated 5 cm.
Step 6: Set one container outside the lower end of the box.
Step 7: Place the end of the tubing in the container. Use a clothespin to clamp the tubing closed just outside the box.
Step 8: Fill the other container with water.
Step 9: Put the tubing from the higher end of the box into the container with the water.
Step 10: Place the container higher than the box.
Step 11: Allow the water to flow through the tubing to the lower end of the box.
Step 12: After 30 minutes, remove the clothespin and allow the water to flow into the second container. Record the temperature of the water in the second container in the data table below. Repeat the experiment two more times.

Variables

Independent: color of tubing
Dependent: temperature of water in tubing
Constants: same box, tap water, length of hose, length of time water is in tubing

Experimental Setup:

Repeat steps 2 through 12 using the white tubing and then the black tubing.

Results: Record water temperatures in the data table below.

Water Temperature in Celsius				
Tubing	**Trial #1**	**Trial #2**	**Trial #3**	**Average**
clear tubing				
white tubing				
black tubing				

Name: ______________________ Date: ______________

Analysis: Study the results of your experiment. Create a graph that will compare the average temperature of the water in each of the three different-colored tubes. Place the dependent variable (temperature) on the *y*-axis. Place the independent variables (color of tubing) on the *x*-axis.

Effects of Colored Tubing on Water Temperature

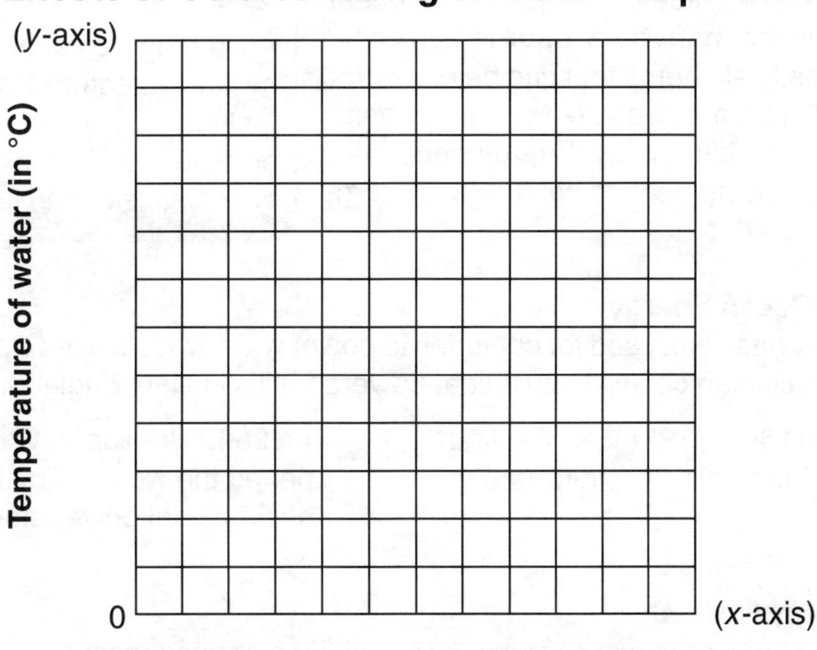

Color of Tubing

Conclusion: Write a summary of the experiment (what actually happened). It should include the purpose, a brief description of the procedure, and whether or not the hypothesis was supported by the data collected. Use key facts from your research to help explain the results. The conclusion should be written in first person ("I").

Ocean Energy
Student Information

Ocean Energy

Ocean waves are created by the wind blowing over its surface and by tides. The power of ocean waves creates an **inexhaustible source** (unlimited resource) of mechanical energy that can be converted into electricity. Different methods are continually being developed to harness this energy. Energy generation devices are already being used to capture the energy of ocean waves and tides.

Science by the Numbers

According to analysis conducted by the Electric Power Research Institute, ocean energy could provide up to 10 percent of the nation's electrical needs.

Early History of Ocean Energy

Tidal energy has been used for centuries to power waterwheels used by mills for grinding grains into flour. American colonists built tidal-powered mills in New England.

- Issac Newton (1642–1727) was the first person to scientifically explain ocean tides.
- In 1966, President Charles de Gaulle opened the world's first and still the most powerful tidal power station in France.

How Ocean Energy is Used

The constant movement of ocean waves and tides creates mechanical energy that can be converted to electrical energy. Wave-powered devices are being used to extract energy from surface waves or from pressure fluctuations beneath the water's surface. Ocean Thermal Energy Conversion (OTEC) is another energy source that uses temperature differences in the ocean's waters to create electricity.

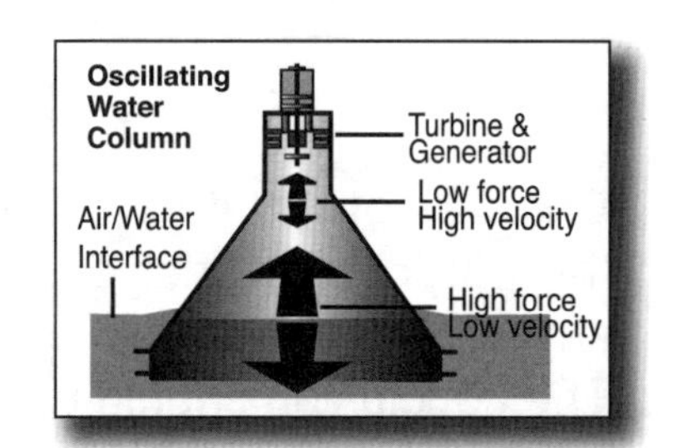

Collecting Wave Energy

1. An **oscillating water column** is a partially submerged column at the ocean's edge with a turbine on top. There is water in the bottom of the column and air in the middle. When the water level rises, it pushes air out the top of the column, through the turbine. Electricity is generated. When the water level recedes, the air is forced back through the turbines in the opposite direction, creating a continuous flow of electricity.
2. A **tapchan** is a tapered channel. This tapered channel feeds into a reservoir that has been built into cliffs that rise above sea level. As the channel narrows, the wave's height increases, causing the water to flow over the walls of the reservoir. This water is fed through a turbine that generates electricity.
3. A **pendulor device** is a rectangular box that has one end open to the sea. A hinged flap (door) connected to a hydraulic pump and generator swings back and forth as the waves move in and out. This door motion causes the generator to create electricity.

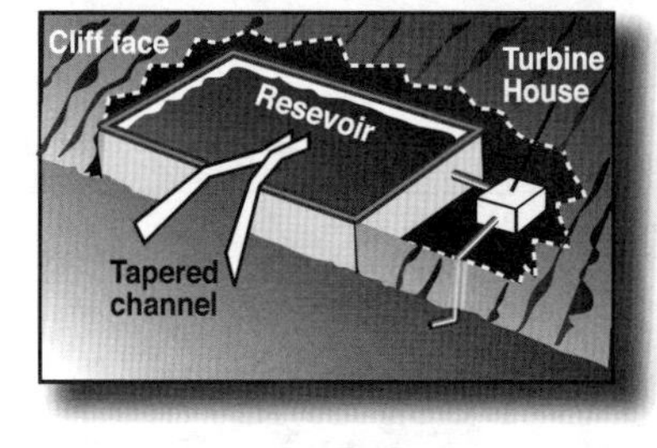

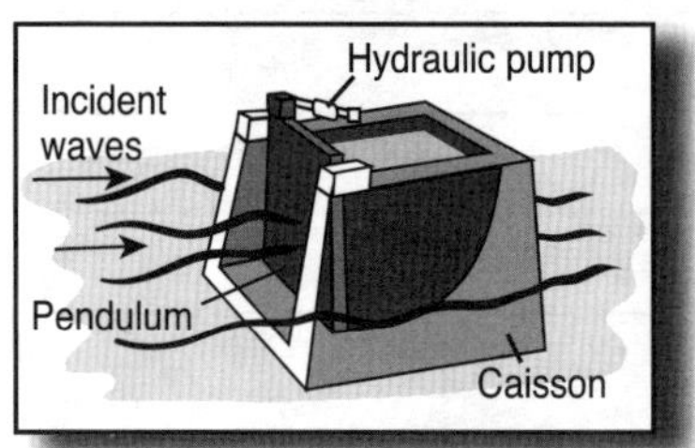

Just the Facts

Tides change from high to low every six hours.

There are no tidal plants operating in the United States.

Water is approximately 800 times denser than air. Therefore, tidal turbines have to be built much sturdier than wind turbines.

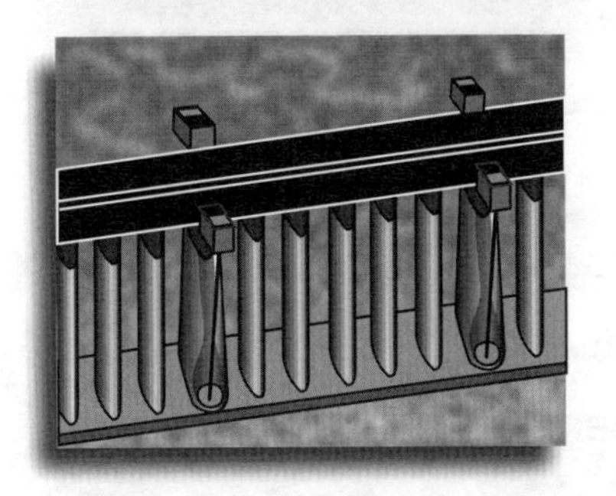

Tidal Energy

Tidal energy is a nonpolluting, inexhaustible source of energy from the rise and fall of tides in the ocean. This energy can be used to generate electricity. The gravitational pull between Earth and the moon causes the tides. The rise and fall in sea level with respect to the land is called **tides**. When the water is at its highest level, we say there is a **high tide**. When the water is at its lowest level, there is a **low tide**. Tidal energy can only be generated in a few places where the difference in high and low tides is at least five meters.

Collecting Tidal Energy

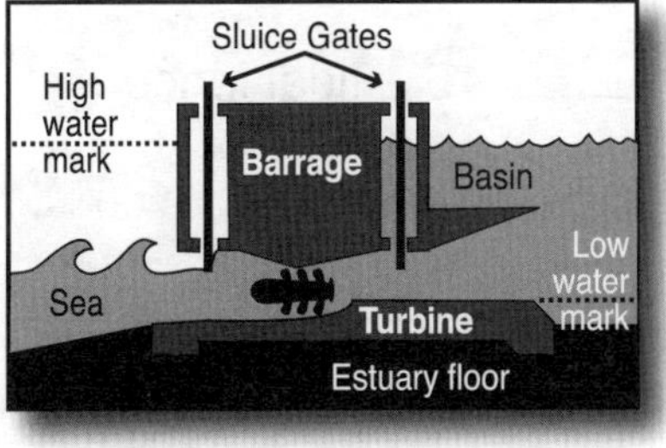
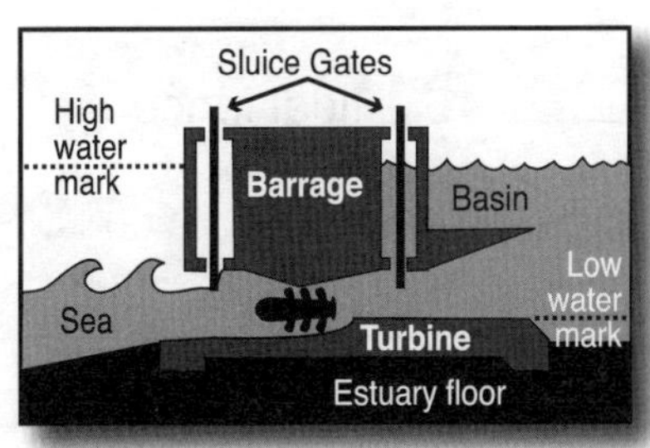

A **tidal barrage** is a huge dam-like structure built across a harbor or narrow inlet. Turbines in the barrage change the moving water into electricity as the tides come in and out. When the tide rises, the water fills the estuary and pushes through the turbines. It then becomes trapped behind the dam. When the tide falls, the water passes through the opposite direction, turning the turbines again.

A **tidal fence** consists of vertical axis turbines placed in a fence. As the water passes, it is forced through the turbines, thereby generating electricity. Tidal fences are used across channels or between two landmasses.

Tidal turbines resemble wind turbines but are built much stronger. They are placed in areas where the tide flow is strong.

Ocean Thermal Energy Conversion

Heat energy stored in the Earth's oceans can be used to generate electricity. The process is called **Ocean Thermal Energy Conversion** (OTEC). The only places in the world where conditions exist for this process to work are located between the Tropic of Capricorn and the Tropic of Cancer.

Future Use of Ocean Energy

As technology develops and affordable, reliable devices become available, more wave energy farms will be seen in the near future.

Advantages

- Tidal energy is a nonpolluting, inexhaustible (renewable) energy source.

Disadvantages

- Tidal fences can disturb large marine animals' movements.
- Not many places have dramatic enough tidal change to support a tidal power plant.

Name: ______________________________ Date: ____________________

Check Point

Matching

_____	1. tidal barrage	a. consists of vertical axis turbines placed in a fence across a channel
_____	2. tapchan	b. when the water is at its highest level
_____	3. high tide	c. a tapered channel
_____	4. tidal turbines	d. a huge dam-like structure built across a harbor or narrow inlet
_____	5. tidal fence	e. resemble wind turbines

Fill in the Blank

6. ______________ ______________ is a nonpolluting, inexhaustible source of energy from the rise and fall of tides in the ocean.
7. The gravitational pull between Earth and the moon causes the ______________.
8. The constant movement of ocean waves and tides creates ______________ energy that can be converted to ______________ energy.
9. The power of ocean waves creates an ______________ ______________ (unlimited resource) of mechanical energy that can be converted into electricity.
10. Heat energy stored in the Earth's oceans can be used to generate electricity through a process called ______________ ______________ ______________ ______________ (OTEC).

Multiple Choice

11. Where was the world's first and still the most powerful tidal power station built?
    a. Italy
    b. Germany
    c. Great Britain
    d. France

12. He was the first person to scientifically explain ocean tides.
    a. Issac Newton
    b. Albert Einstein
    c. Benjamin Franklin
    d. Joseph Henry

Constructed Response

Explain why scientists consider tidal energy an inexhaustible resource. Give specific examples and details to support your answer. Answer on your own paper.

Name: ______________________ Date: ______________

Mini Labs

Mini Lab #1: Wave Action

Materials:

blue food coloring
1 cup mineral oil
2 cups water
small funnel
1 clear plastic 2-liter bottle (labels removed)
1 lid for plastic bottle

Procedure: Pour the water into the bottle using the funnel. Add food coloring and oil to the water, again using the funnel. Cap bottle tightly and turn bottle horizontally. Now raise and lower the ends of the bottle.

Observation: Describe the action of the bottled wave. Does the wave move through the water, or does the wave make the water move along with it?

____

____

Conclusion: What can you conclude about the energy of a wave? ______________

____

____

Mini Lab #2: Wind and Waves

Materials:

rectangular cake pan
water
electric fan

Procedure: Place the cake pan on a table. Fill the pan three-fourths full with water. Place the fan 30 cm from the narrow end of the pan. Turn the fan on low speed and have it blow across the water. Observe the top of the water. Next, turn the fan on medium speed and then high. Observe the top of the water.

Observation: What effect did the fan have on the surface of the water? ______________

____

____

Conclusion: What is the connection between the wind and the waves? ______________

____

____

Name: ______________________________ Date: ____________________

Inquiry Lab: Energy From the Deep

Purpose: The purpose is a question that asks what you want to learn from the investigation. It should be clearly written, it usually starts with the verb "Does," and it can be answered by measuring something.

Purpose: *Does water depth affect wave speed?*

Research: The goal of the research is to find information that will help you make a prediction about what will occur in your experiment. Investigate ocean energy, ocean waves, and ocean currents. Use the lines below for note taking.

____

____

____

____

____

____

____

____

____

____

____

____

____

Online Science: Learn more about ocean energy at the following website. "Hydropower Basics—Energy from Moving Water: Tidal Power." U.S. Department of Energy. <http://www.eia.doe.gov/kids/energyfacts/sources/renewable/hydropower.html>

Hypothesis: Make an educated guess about what you think will happen in your project. Your hypothesis should be clearly written. It should answer the question stated in the purpose, be brief and to the point, and it should identify the independent and dependent variables.

Example: *The depth of water* (choose one) ***will** or **will not** affect wave speed.*

Hypothesis: ____

____

____

Name: ______________________ Date: ______________

Procedure: The procedure is a plan for your experiment. The plan includes a list of the materials needed, step-by-step how-to directions (written like a recipe) for conducting the experiment, and it identifies the variables. Measurements are made and recorded using metric units.

Materials Needed:

stream tray — water

1 long block of wood — 1 short block of wood

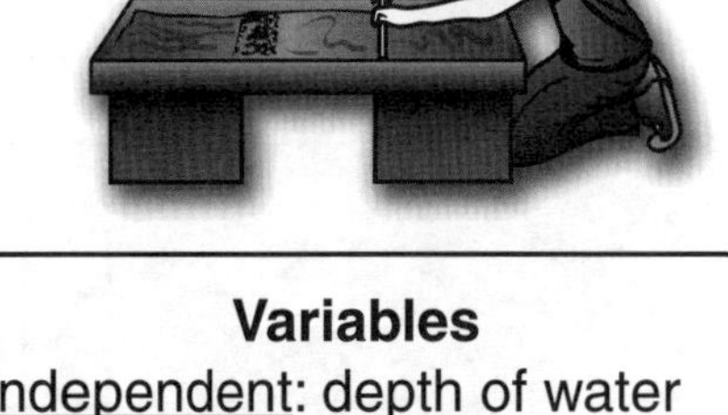

Experiment:

Controlled Setup:

Step 1: Fill the stream tray to a depth of 1 cm. Use the longer piece of wood to act as a reflector at one end of the tank.

Step 2: Practice making a wave pulse by pushing the shorter piece of wood into the water at one end. Don't make a wave too big or you will have problems with the depth.

Variables
Independent: depth of water
Dependent: wave speed
Constants: same stream tray, same block of wood

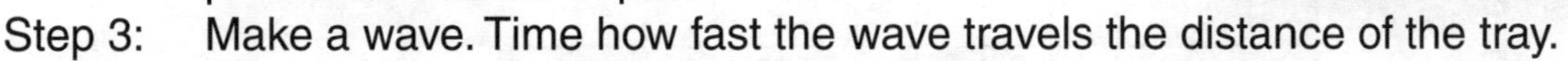

Step 3: Make a wave. Time how fast the wave travels the distance of the tray.

Step 4: Record the distance and time in the data table below.

Step 5: Repeat steps 3 and 4 two more times.

Step 6: Calculate the average time. Then, calculate the speed for each test and the average speed by dividing the distance by the time.

Experimental Setup:

Repeat Steps 3 through 6 for water depths of 2 cm and 4 cm.

Results:

Controlled Setup			
Depth (cm)	**Distance**	**Time (sec)**	**Speed (m/sec)**
1 cm		Test #1	
		Test #2	
		Test #3	
		Average	
Experimental Setup			
Depth (cm)	**Distance**	**Time (sec)**	**Speed (m/sec)**
2 cm		Test #1	
		Test #2	
		Test #3	
		Average	
4 cm		Test #1	
		Test #2	
		Test #3	
		Average	

Name: ______________________ Date: ______________

Analysis: Study the results of your experiment. Create a graph that will compare the effect of water depth on wave speed in the control group with the experimental group. Place the dependent variable (speed) on the *y*-axis. Place the independent variable (depth of water) on the *x*-axis.

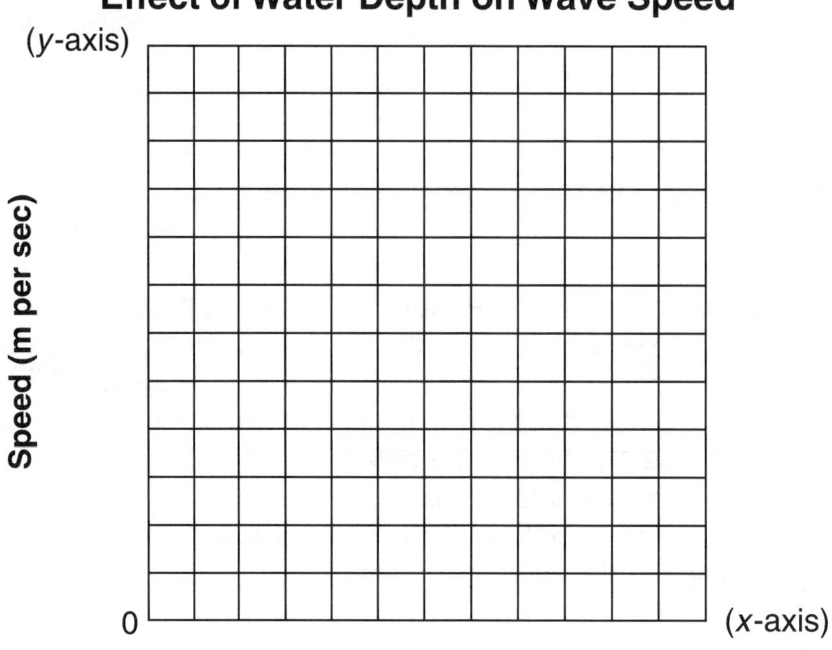

Depth of Water (cm)

Conclusion: Write a summary of the experiment (what actually happened). It should include the purpose, a brief description of the procedure, and whether or not the hypothesis was supported by the data collected. Use key facts from your research to help explain the results. The conclusion should be written in first person ("I").

Biomass Energy
Student Information

Plants and Animals

Biomass is plant material and animal waste used as fuel. Biomass can be thought of as stored energy from the sun. During **photosynthesis** (the process plants use to make food), plants absorb light energy from the sun. They use the energy to produce carbohydrates or chemical energy, which is stored in the plant as sugar, starches, and cellulose. The chemical energy in plants gets passed on to animals when they eat plants or eat other animals who have eaten plants.

Science by the Numbers

Three gallons of ethanol can be produced from one bushel of corn.

According to the United States Department of Energy, biofuels provide about 3% of the energy used in our country.

Biomass is a renewable energy source used to produce biofuel. Plant and animal byproducts used to produce biofuel can be grown over and over. Biomass includes wood and wood wastes, trash from landfills, and crop and animal waste. Burning biomass releases the stored chemical energy as heat. Biomass can also be converted to other forms of energy, such as methane gas, ethanol (grain alcohol), and biodiesel.

Early History of Biomass Energy

- Wood was the first form of biofuel. Ancient people used it for cooking and heating.
- In the 1600s, alcohol stoves (also called spirit lamps) were commonly used by travelers to warm food and themselves.
- In the early 1800s, lamp fuels in the United States and Europe were usually made from animal and vegetable oils.
- In the 1890s, Rudolph Diesel invented an engine that ran on peanut oil.
- George Washington Carver (1864–1943), an agricultural researcher, explored the many uses of peanuts.
- In 1903, Henry Ford designed the Model T car to run on ethanol.
- From 1939 to 1945, many countries mixed grain alcohol with gasoline because World War II had caused a fuel shortage.

How Biomass is Used

Wood products, hay, and corn can be burned in biomass furnaces and stoves for home heating. Biomass can also be used much like fossil fuels to generate electricity in power plants. Biofuels are being used around the world as an alternative to fossil fuels. Biofuels are currently being used in transportation. Ethanol is an alcohol produced from crops, such as corn and sugar cane. It is combined with gasoline for use in any gasoline vehicle in the United States. Scientists are experimenting with making ethanol from cellulose, or plant fiber, such as leaves and stems. Biodiesel produced from natural oils, such as soybean oil, can safely be used in any vehicle with a diesel engine.

Just the Facts

Brazil, the largest sugar-producing nation, is also the world's leading exporter of ethanol.

Corn is the leading crop used to produce ethanol fuel for transportation in the United States.

The San Francisco Zoo composts 20,000 cubic yards of herbivore animal waste and 200 tons of green matter every year.

Almost all the ethanol produced in the United States is used to make a mixture of ethanol and gasoline called "gasohol" or "E10". Fuel with 85% ethanol, called "E85", is available in limited areas.

Sources of Biomass Energy

Plants: The sugar found in grains and other plants can be fermented to make an alcohol fuel called **ethanol**. Most ethanol used in the United States today is distilled from corn. Other plants used to make ethanol include wheat, potatoes, rice, sugar cane, and sugar beets.

Plant Oil and Animal Fat: Plant oil, animal fat, or grease can be used to make **biodiesel**. Most biodiesel is made from soybeans.

Animal Waste: Animal waste or manure can be used to produce a gas called **methane**. Some large livestock farms such as cattle, hog, and poultry operations use a machine called a methane digester to separate the methane from the animal waste. The methane can then be used to generate electricity for the farm or sold to electric companies.

Garbage: Methane gas can also be produced by decaying garbage at landfill sites. Pipes can be driven deep into the decaying garbage to capture the gas. The gas can then be collected, treated, and sold by the landfills or used to heat buildings or generate electricity.

Future Use of Biofuels

In 1970, the stage was set for developing cleaner-burning fuels with the passage of the Clean Air Act by the Environmental Protection Agency (EPA). In 2009, the passage of the American Recovery and Reinvestment Act provided money for the research and development of biofuels. Scientists are trying to convert plant material, such as corn stalks and husks, switchgrass, and wood chips, into fuel that can power cars and trucks.

Advantages

- Biomass is renewable, efficient, and clean burning.
- People can use whatever biomass is available.
- Biofuels are considered **carbon neutral**. The fuels emit the same amount of carbon dioxide when burned as the amount absorbed by the plants used to make the fuel.

Disadvantages

- Planting biofuel crops instead of food crops could drive up food prices.
- Ethanol is expensive to produce.

Name: ______________________________ Date: ____________________

Check Point

Matching

_____ 1. biomass
_____ 2. biodiesel
_____ 3. photosynthesis
_____ 4. ethanol
_____ 5. gasohol

a. the process plants use to make food
b. an alcohol fuel produced from corn
c. a mixture of ethanol and gasoline
d. made from plant oil, animal fat, or grease
e. plant material and animal waste used as fuel

Fill in the Blank

6. ____________________, the largest sugar-producing nation, is also the world's leading exporter of ethanol.
7. ____________________ are considered carbon neutral.
8. Methane gas is produced by ____________________ garbage at landfill sites.
9. Most biodiesel is made from ____________________.
10. Biofuels are being used around the world as an alternative to ____________________ ____________________.

Multiple Choice

11. Which is the leading crop used to produce ethanol fuel in the United States?
    a. soybeans b. wheat
    c. corn d. beets

12. How many gallons of ethanol can be produced from one bushel of corn?
    a. 1 b. 4
    c. 7 d. 3

Constructed Response

Explain why scientists consider biofuels carbon neutral. Give specific examples and details to support your answer.

____
____
____
____
____
____

Name: ______________________________ Date: ____________________

Mini Labs

Mini Lab #1: Fermentation

Materials:

1 packet of dry yeast
1 plastic liter bottle
1 lid for plastic bottle
1 9-inch balloon
50 grams sugar
250 mL warm water

Procedure: Pour the warm water into the bottle. Add sugar to the water. Cap the bottle and shake until the sugar is dissolved. Add yeast to the sugar solution in the bottle. Place the balloon over the mouth of the bottle. Observe the balloon over the course of a day.

Observation: What do you notice happening to the balloon? ______________________

____

Conclusion: How is the process of fermentation used to produce ethanol? ____________

____

____

Mini Lab #2: Landfill

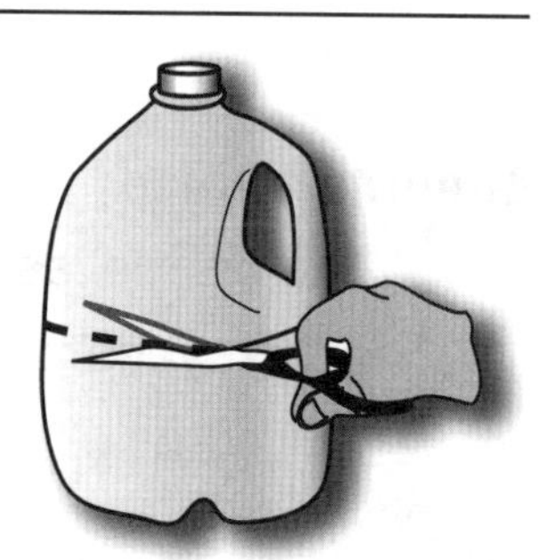

Materials:

1 gallon plastic milk jug
clear plastic wrap
rolling pin
modeling clay
1 foot of glass tubing
balloon
organic compost
scissors
rubber band

Procedure: Cut the top from a gallon plastic milk jug, leaving the sides of the bottom about 10 –12 cm high (throw the top away). Fill the jug bottom half full of compost. Thoroughly wet the compost. Cover the opening of the jug with clear plastic wrap. Insert the glass tubing through the plastic wrap and into the compost. Roll out the modeling clay into a thin layer. Cover the plastic wrap with the layer of modeling clay. Seal the edges of the jug and around the glass tubing with the clay also. Place the opening of the balloon over the glass tubing. Attach the balloon to the tube with a rubber band. Observe your milk jug landfill for two weeks.

Observations: What do you notice happening to the balloon after a couple of weeks?

____

Conclusion: Describe what is happening in your milk jug landfill. ________________

____

____

Name: ______________________ Date: ______________

Inquiry Lab: Energy From Peanuts

CAUTION!—HEALTH HAZARD: This procedure involves peanuts. If any students are allergic to nuts, they should not remain in the classroom and should be excused from this lab. Inform students in other classrooms or adjoining science laboratories or students who may use the same lab for other classes.

Purpose: The purpose is a question that asks what you want to learn from the investigation. It should be clearly written, it usually starts with the verb "Does," and it can be answered by measuring something.

Purpose: *Does the type of nut burned affect the number of calories of heat produced?*

Research: The goal of the research is to find information that will help you make a prediction about what will occur in your experiment. Investigate biomass, nuts, and calories. Use the lines below for note taking.

____

____

____

____

____

____

____

____

____

____

Online Science: Learn more about biomass at the following website. "Biomass Energy—Renewable Energy from Plants and Animals." <http://www.eia.doe.gov/kids/energyfacts/sources/renewable/biomass.html>

Hypothesis: Make an educated guess about what you think will happen in your project. Your hypothesis should be clearly written. It should answer the question stated in the purpose, be brief and to the point, and identify the independent and dependent variables.

Example: *The type of nut burned* (choose one) ***will*** *or* ***will not*** *affect the number of calories of heat produced.*

Hypothesis: ____

____

Name: ______________________________ Date: ____________________

Procedure: The procedure is a plan for your experiment. The plan includes a list of the materials needed, step-by-step how-to directions (written like a recipe) for conducting the experiment, and it identifies the variables. Measurements are made and recorded using metric units.

Materials Needed:

large test tube
graduated cylinder
room temperature water
laboratory balance
holder ring stand
needle
thermometer
safety goggles
cork
matches
tube holder
peanuts, cashews, and walnuts

Experiment: (Caution: This Lab needs close teacher supervision.)

Controlled Setup:

Step 1: Put on your safety goggles.

Step 2: Using the balance, find the mass of the peanut. Record the value in the data table.

Step 3: Set up the ring stand and secure the test tube in the test tube holder.

Step 4: Using the graduated cylinder, measure 10 mL of water. Pour the water into the test tube. Measure the temperature of the test tube water in degrees Celsius. Record this value as the starting temperature in the data table.

Variables

Independent: type of nut
Dependent: number of calories produced
Constants: same test tube and nut holder, distance from bottom of test tube, room temperature water, amount of water

Step 5: Carefully push the eye of the needle into the center of the smaller end of the cork. Then gently push the pointed end of the needle into a peanut. (If it breaks, try another peanut.) Place the test tube to 2 cm above the nut.

Step 6: Remove the peanut and holder from under the test tube. Light the peanut. Once the peanut has started burning, position it directly beneath the water-filled test tube.

Step 7: When the peanut has stopped burning, check the temperature of the water. Record this value in the data table.

Step 8: Pour out the test tube water.

Step 9: Place the burned peanut on the balance and determine its end mass. Record this value in the data table.

Experimental Setup:

Repeat steps 1 through 9 using the cashews and walnuts.

Results: Record test results in the data table below. To calculate the calories per gram of the nut, use the following equation:

Calories per gram = (heat gained by water)/(mass lost when nut burned)

Nut	Beginning Mass (g)	Ending Mass (g)	Starting Temp (°C)	Final Temp (°C)	Heat Gained	Calories per gram
Peanut						
Cashew						
Walnut						

Name: ______________________________ Date: ____________________

Analysis: Study the results of your experiment. Create a graph that will compare the number of calories per gram produced by each kind of nut. Place the dependent variable (calories per gram) on the *y*-axis. Place the independent variable (type of nut) on the *x*-axis.

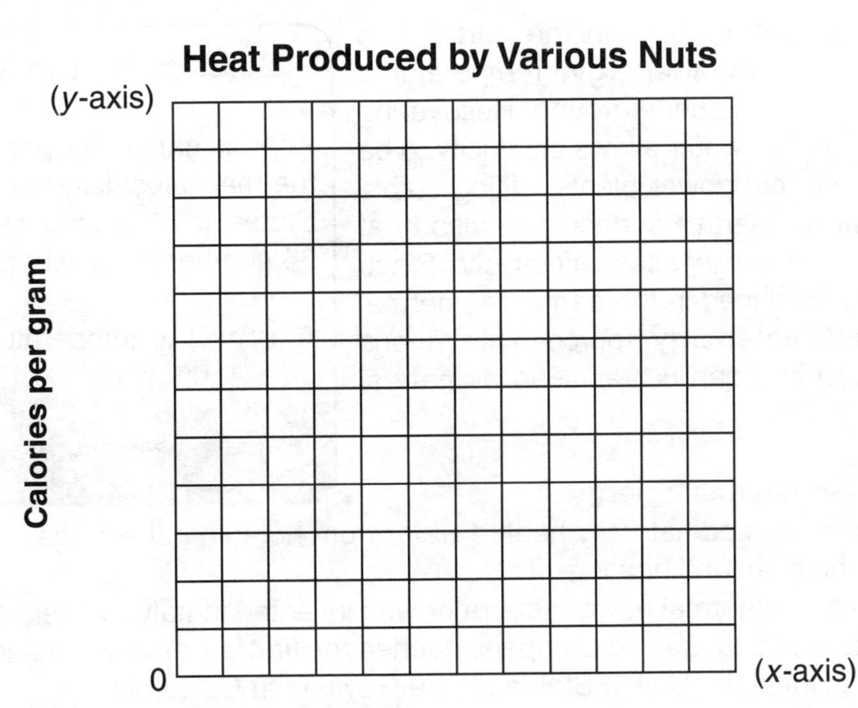

Conclusion: Write a summary of the experiment (what actually happened). It should include the purpose, a brief description of the procedure, and whether or not the hypothesis was supported by the data collected. Use key facts from your research to help explain the results. The conclusion should be written in first person ("I").

____

____

____

____

____

____

____

____

____

____

____

____

Geothermal Energy
Student Information

Heat energy is stored within the earth. This energy is called **geothermal energy**. Temperatures increase the deeper you go underground. Recovering this heat as steam or hot water allows electricity to be generated in geothermal power plants. Using stable temperatures near the earth's surface can also heat and cool homes and businesses effectively. Since heat is continually produced in the earth, geothermal energy is a **renewable energy** source. This means that it is not limited and can be replaced by natural processes.

Science by the Numbers

Approximately 90 percent of geothermal electricity in the United States is produced in California's 34 geothermal power plants.

A constant temperature between 50°F (10°C) and 60°F (16°C) can be found 5 to 10 feet below the earth's surface.

Early History of Geothermal Energy

Ancient cultures used naturally heated water from hot mineral springs. This water was used for cooking, bathing, and heating.

- In 1904, the first geothermal power generator was invented in Italy by Piero Conti.
- In 1960, Pacific Gas and Electric Company opened the first large-scale geothermal electricity-generating plant in the United States at The Geysers in California.

How Geothermal Energy is Used

Geothermal heat is used two ways. The first way uses **thermal** (heat) **energy** stored in the earth that is converted into electrical energy. The second way uses ground heat pumps that use the constant temperature stored near the earth's surface to heat and cool buildings.

Geothermal Reservoirs: Hot water and steam can be trapped under pressure in cracks and pockets called geothermal reservoirs. Water in these reservoirs can reach temperatures of 700°F (371°C). Geothermal power plants are built in areas where the reservoirs are close to the surface (1 to 2 miles). By drilling into the reservoir, steam or hot water can be piped into power plants, and this heat can be converted into electricity.

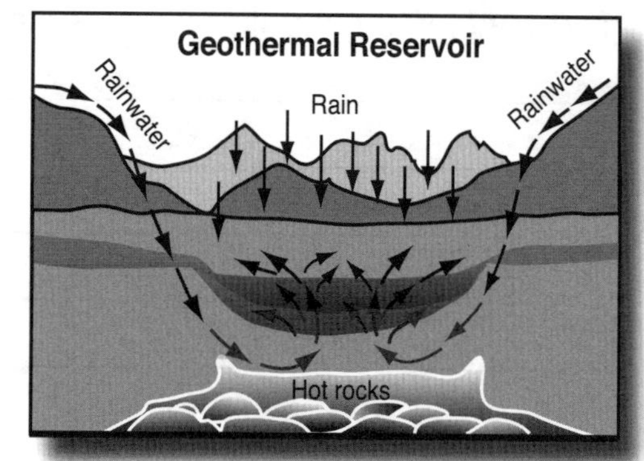

Geothermal Heat Pumps: A heat pump can be used almost anywhere without a geothermal reservoir. In most areas, soil temperatures are warmer than the air in the winter and cooler than the air in summer. A heat pump consists of a water-filled loop of plastic pipe that is buried underground where temperature is constant. During winter, the water passes through the pipe, picking up heat from the ground. The heat is then released into the building, making it several degrees warmer. In the summer, air is warmer than the underground temperature. The warm air in the building is cooled by the water from the underground pipes; it then re-enters the building, providing air conditioning.

Just the Facts

The United States generates more geothermal electricity than any other country.

The U.S. Environmental Protection Agency (EPA) has rated geothermal heat pumps as the most energy efficient, environmentally clean, and cost-effective systems for heating and cooling.

Today, people still bathe in hot springs. Some believe the minerals found in hot springs have healing powers.

In Reykjavik, Iceland, almost all of the buildings are heated by geothermal energy.

Collecting Geothermal Energy (Power Plants)

Dry steam plants are rare. They use steam (no water) that shoots up directly from a geothermal reservoir to turn generator turbines.

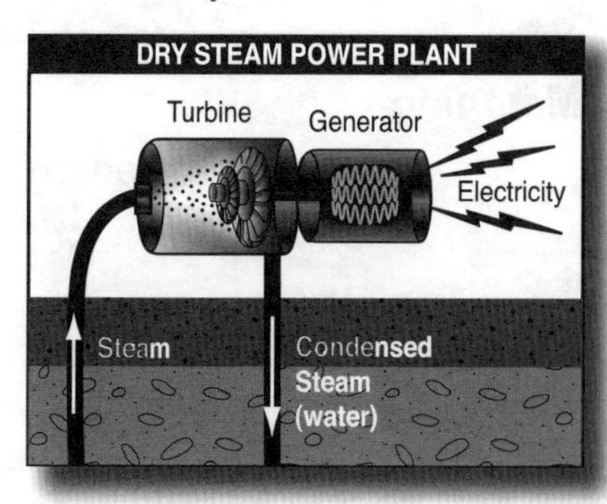

Flash steam plants are the most common geothermal power plant. Flash steam plants release the high-pressure hot water from the deep reservoir in a flash tank. It is then converted into steam that drives the generator turbines. When the steam cools, the unused geothermal water and the condensed steam are injected back into the ground so it can be used over and over again. New Zealand invented the first flash technology.

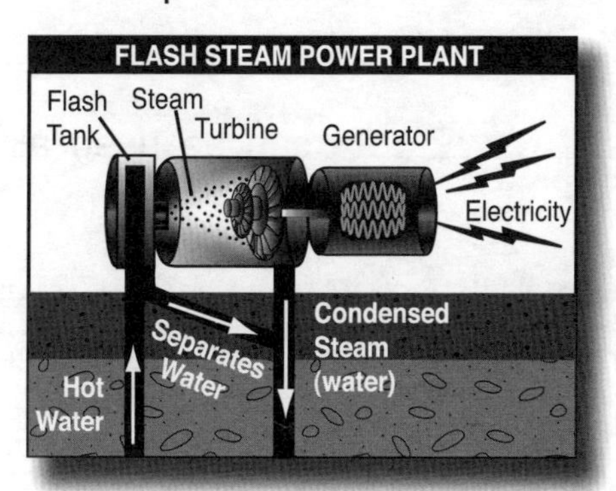

Binary cycle power plants transfer the heat from geothermal water to a second liquid in separate adjoining pipes. This liquid turns to steam and is used to power the turbine generator. Lower temperature reservoirs can be used with binary technology; therefore a greater number of geothermal reservoirs can be used.

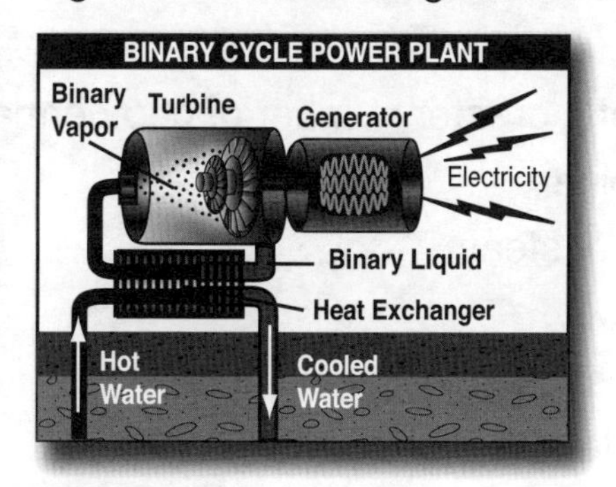

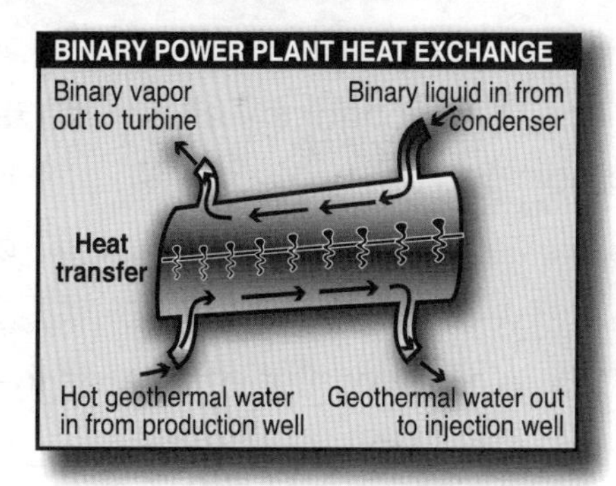

Future Use of Geothermal Energy

There is a bright future for geothermal energy. It is expected that within ten years, geothermal energy capacity will grow. Already, thousands of homes and buildings across the United States are using geothermal heat pumps for heating and cooling.

Advantages

- Geothermal energy is a clean, safe, renewable energy source.
- Geothermal energy has no emissions from burning fuels.

Disadvantages

- Geothermal power plants can only be built in areas where geothermal reservoirs are close to the earth's surface.

Name: ______________________ Date: ______________

Check Point

Matching

_____	1. renewable energy	a.	heat energy
_____	2. dry steam plant	b.	most common geothermal power plant
_____	3. thermal energy	c.	heat energy stored within the earth
_____	4. flash steam plant	d.	uses steam that shoots up directly from a geothermal reservoir
_____	5. geothermal energy	e.	can be replaced by natural processes

Fill in the Blank

6. A ____________ ____________ can be used almost anywhere without a geothermal reservoir.
7. ____________ ____________ ____________ ____________ transfer the heat from geothermal water to a second liquid in separate adjoining pipes.
8. The first geothermal power plant was invented in ____________ by Piero Conti in 1904.
9. ____________ ____________ invented the first flash technology.
10. Approximately 90 percent of geothermal electricity in the United States is produced in the 34 geothermal power plants in ____________.

Multiple Choice

11. This technology allows the use of a greater number of geothermal reservoirs.
    a. flash
    b. binary
    c. hot spring
    d. dry steam

12. Which country generates the most geothermal electricity?
    a. United States
    b. New Zealand
    c. Italy
    d. Iceland

Constructed Response

Explain why scientists consider geothermal energy an inexhaustible resource. Give specific examples and details to support your answer.

____

____

____

____

Name: ______________________ Date: ______________

Mini Labs

Mini Lab #1: Which Is Cleaner?

Materials:

stove top
small sauce pan
1 cup water
tongs
2 sheets of plain paper
household candle
matches

Procedure: Heat the water in the small pan on the stove top. When the water turns to steam, grasp a piece of paper with the tongs and hold it approximately 15 to 16 cm above the steam. Do this for 30 seconds, and then set aside. Next, light the candle. Hold the other sheet of paper with the tongs over the candle's flame at the same height. Wait 30 seconds, blow out the candle, and examine both sheets of paper.

Observation: Do you notice a difference between the two sheets of paper? If so, what is the difference? ______________________

____

Conclusion: ______________________

____

Mini Lab #2: Underground Geyser Warmth

Materials:

large mixing bowl
large nail
drinking straw
water
safety goggles
food coloring
hammer
modeling clay
sauce pan
small glass bottle with screw cap
stove top for heating water
pin
oven mitts

Procedure: Punch a hole in the bottle cap using the hammer and nail. Fill the sauce pan with water and bring it to boiling point on the stove top. Now, half fill the glass bottle with cool water. Add 4 to 5 drops of food coloring. Tighten the cap and insert the straw through the hole in the cap downward into the water. Seal the hole around the cap and straw with the clay. Place a small bit of clay in the top of the straw. Using the pin, make a tiny hole that goes all the way through the clay. Pour the boiling water into the mixing bowl. Stand the bottle in the bowl.

Conclusion: Based on your observations, how are geysers produced?

____

____

____

Name: ______________________________ Date: ____________________

Inquiry Lab: Geothermal Steam Power

Purpose: The purpose is a question that asks what you want to learn from the investigation. It should be clearly written, it usually starts with the verb "Does," and it can be answered by measuring something.

Purpose: *Does the number of steam release holes affect the spin of a pinwheel?*

Research: The goal of the research is to find information that will help you make a prediction about what will occur in your experiment. Investigate geysers, geothermal energy, and steam power. Use the lines below for note taking.

____

____

____

____

____

____

____

____

____

____

____

____

____

Online Science: Learn more about geothermal energy at the following website. "Geothermal Basics—What Is Geothermal Energy?." U.S. Department of Energy. <http://tonto.eia.doe.gov/kids/energy.cfm?page=geothermal_home-basics>.

Hypothesis: Make an educated guess about what you think will happen in your project. Your hypothesis should be clearly written. It should answer the question stated in the purpose, be brief and to the point, and identify the independent and dependent variables.

Example: *The number of steam release holes* (choose one) ***will*** *or* ***will not*** *affect the spin of a pinwheel.*

Hypothesis: ____

____

Name: ______________________________ Date: ____________________

Procedure: The procedure is a plan for your experiment. The plan includes a list of the materials needed, step-by-step how-to directions (written like a recipe) for conducting the experiment, and it identifies the variables. Measurements are made and recorded using metric units.

Materials Needed:

child's pinwheel
aluminum foil
water
10 p nail
oven mitt
timer
hot plate or stove top
1 3-qt. cooking pot
black marker

Variables
Independent: number of holes
Dependent: number of pinwheel spins
Constants: same pan, amount of water, height of pinwheel above hole in foil

Experiment:

Controlled Setup:

Step 1: Pour 2 L of water in the pan. Make a foil lid for the pan by covering the top with two layers of aluminum foil. Crimp edges for a tight seal.

Step 2: Punch a small hole in the center of the two sheets of foil using the nail.

Step 3: Place the pan on the hot plate (medium heat) and bring to a boil. Do not let boiling water touch the bottom of the foil.

Step 4: Mark a large X on the back of the pinwheel on one of the blades.

Step 5: Boil the water for several minutes. When the steam comes out the hole and the foil cover is slightly raised, you are ready to begin your experiment.

Step 6: Hold the pinwheel face down, 30 cm above the hole in the foil.

Step 7: Note the location of the X mark on the pin wheel. When the X mark reaches the location again, count 1 spin. Count the number of spins made in 60 seconds. Record in the data table. Repeat this step two more times and record the data in the table.

Step 8: Remove the pan and let it cool. Carefully, remove foil. Pour the water out of the pan.

Experimental Setup:

Repeat steps 1 through 8 using 5 holes and then 10 holes.

Results: Record the number of spins in the data table. Calculate and record the average number of spins.

Controlled Setup				
Number of Holes	**Trial #1**	**Trial #2**	**Trial #3**	**Average**
1				
Experimental Setup				
Number of Holes	**Trial #1**	**Trial #2**	**Trial #3**	**Average**
5				
10				

Name: ______________________ Date: ______________

Analysis: Study the results of your experiment. Create a graph that will compare the average number of spins made by the pinwheel in the control setup and the experimental setups. Place the dependent variable (number of spins) on the *y*-axis. Place the independent variable (number of holes) on the *x*-axis.

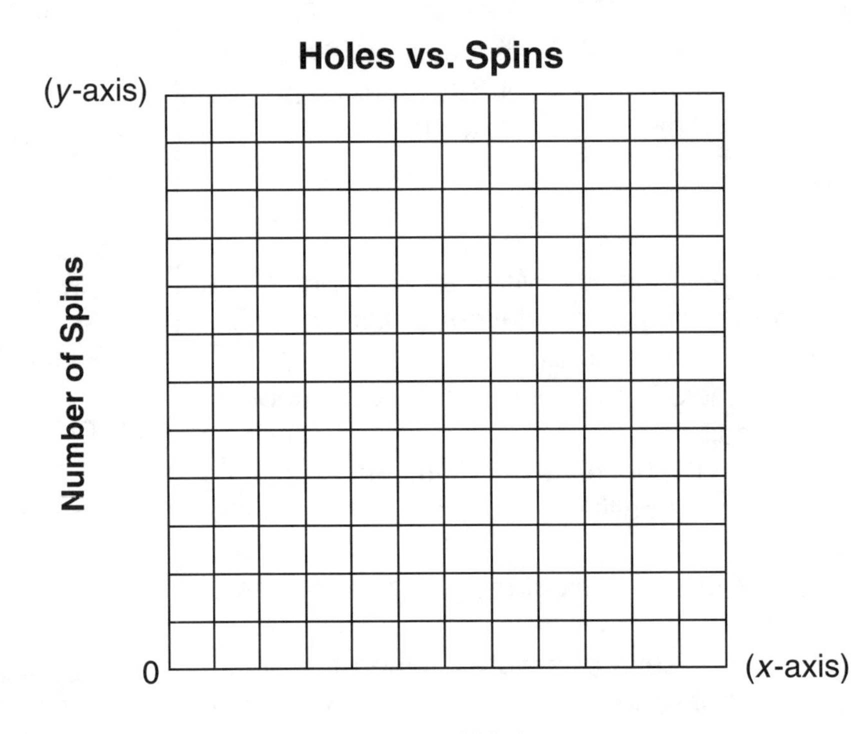

Conclusion: Write a summary of the experiment (what actually happened). It should include the purpose, a brief description of the procedure, and whether or not the hypothesis was supported by the data collected. Use key facts from your research to help explain the results. The conclusion should be written in first person ("I").

____

____

____

____

____

____

____

____

____

____

____

Hydro Energy
Student Information

The amount of water on the earth has remained the same since its formation. The **water cycle** is an exchange of water between land, bodies of water, and the atmosphere. Water moves from land to the atmosphere in the repeated process of evaporation, condensation, and precipitation. This means that water is a **renewable resource** (inexhaustible). We will never run out.

Science by the Numbers

Grand Coulee Dam, the nation's largest hydroelectric power plant, generates approximately 31% of the total U.S hydropower.

Norway produces 99% of its electricity with hydropower.

Moving water has energy. Water is transformed from **potential energy** (stored energy) to **kinetic energy** (moving energy). The mechanical energy of moving water can be converted to electricity. The amount of energy produced is determined by the water's flow or fall. Swiftly flowing water and water plunging from great heights produce the most energy.

Early History of Hydro Energy

Ancient people harnessed the energy of water for farming, simple machines, and transportation.

- The ancient Greeks used waterwheels to turn large stones that ground grain into flour.
- By the 1700s, water was heated to operate steam-powered machines.
- In the 1800s, homesteaders in the United States traveled to water wheel mills to grind their wheat and corn into flour and meal, and water wheels were also used to power sawmills to cut wood.
- In 1882, the first U.S. hydroelectric power plant opened on the Fox River near Appleton, Wisconsin.

How Hydro Energy is Used

Hydro energy is used to produce electricity. **Hydroelectricity**, also known as hydropower, is generated in hydroelectric power stations. Some power plants use dams to produce electricity and some do not. Dams are built on rivers and lakes. This creates reservoirs where the water is stored. Releasing the stored water through big pipes called **penstocks** causes turbines (giant wheels) at the base of the dam to turn. This spins a generator that converts the moving water into electricity. Electricity is then transported to homes and businesses through power lines.

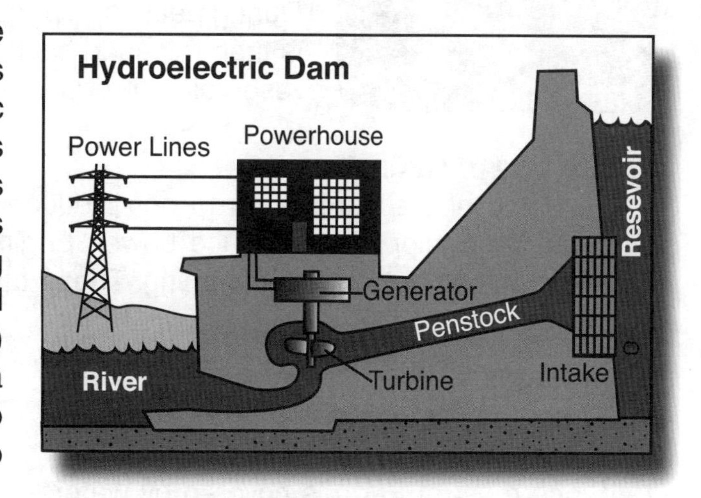

Just the Facts

The first hydro plant built at Niagara Falls in New York was completed in 1879.

Hoover Dam on Lake Mead in Nevada was built from 1931 to 1936. Its power plant produces enough electricity to power a city of 750,000 people.

There are approximately 80,000 dams nation-wide, but most of them do not produce electricity. The main reason dams were constructed was to provide irrigation and flood control.

Hydroelectricity is the largest renewable source of energy on Earth.

Collecting Hydro Energy
Three Basic Types of Hydropower Plants

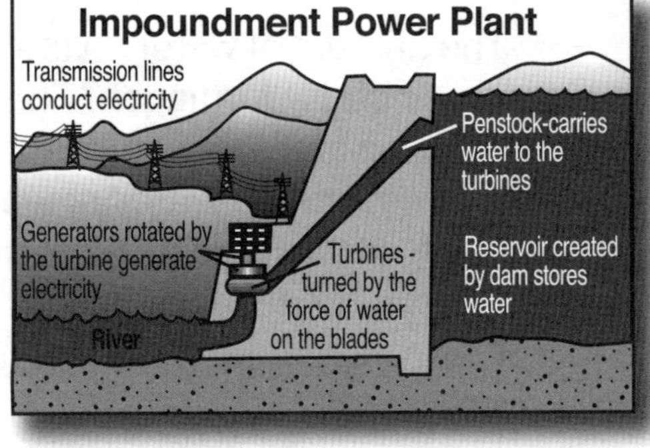

Impoundment Power Plants: Impoundment plants are the most common. Water is stored behind a dam in a reservoir such as a lake. Water is released from the reservoir and fed into a lower-lying power station where it flows through a turbine. The spinning turbine activates a generator that produces electricity.

Run-of-River Power Plants: These are also known as diversion power plants. They use the natural flow of water in a river or canal and may not require a dam. Water is channeled into a canal or penstock and then directed through turbines and a generator, thereby creating electricity.

Pumped Storage Power Plants: These plants use two water reservoirs to store water. When electricity demand is low, like at night, water is pumped from a lower reservoir to an upper reservoir and stored until there is a demand for it. During peak load periods, the water is then released back into the lower reservoir to produce electricity.

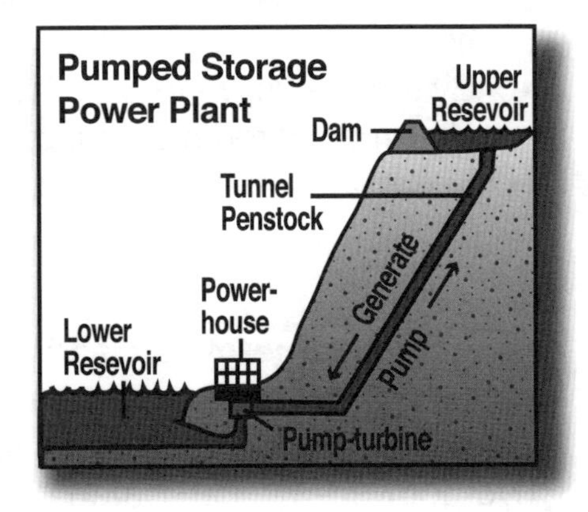

Future Use of Hydro Energy

Many of the dams that do not produce electricity are being studied to determine if they can be equipped to produce power. Engineering research is now focusing on building turbines that do not require damming. These turbines would allow fish to pass through them unharmed.

Advantages

- Water power is a clean (nonpolluting) energy source.
- Since most of Earth is covered by water and cycles through the water cycle, there is an endless supply.

Disadvantages

- Dams and power stations are expensive to build.
- Dams interfere with the life cycles of many fish species, and other wildlife habitat may be destroyed where dams are built.

Name: ______________________ Date: ______________

Check Point

Matching

_____ 1. potential energy		a. hydropower
_____ 2. hydroelectricity		b. large water pipes
_____ 3. renewable resource		c. stored energy
_____ 4. penstocks		d. moving energy
_____ 5. kinetic energy		e. inexhaustible

Fill in the Blank

6. Water moves from land to the atmosphere in the repeated process of ______________, ______________, and ______________.
7. ______________ produces 99% of its electricity with hydropower.
8. ______________ is the largest renewable source of energy on Earth
9. The ______________ energy of moving water can be converted to electricity.
10. Some power plants use ______________ to produce electricity and some do not.

Multiple Choice

11. What is another name for a run-of-river power plant?
    - a. impoundment power plant
    - b. diversion power plant
    - c. pumped storage power plant
    - d. turbine mill

12. This is the most common type of power plant.
    - a. run-of-river
    - b. impoundment power plant
    - c. diversion power plant
    - d. pumped storage plant

Constructed Response

Explain why engineering research is now focusing on building turbines that do not require damming. Give specific examples and details to support your answer.

____

____

____

____

____

____

____

Name: ______________________________ Date: ____________________

Mini Lab

Mini Lab: Demonstrating Hydropower

Materials:

pencil
16 p nail
scissors
9-inch aluminum pie pan
string
ruler
small metal nut
running water

Procedure: CAUTION: This project produces sharp edges. Take appropriate safety measures.

Step 1: Cut the bottom, inner circle out of a 9" aluminum pie pan.

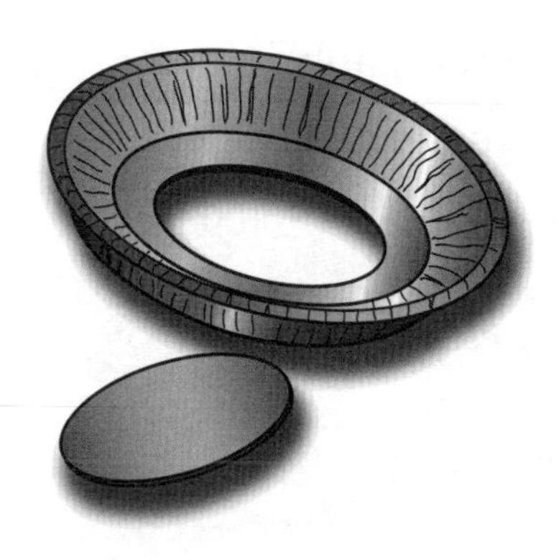

Step 2: Using the ruler, find the center of the circle and punch a hole in the center with the nail.

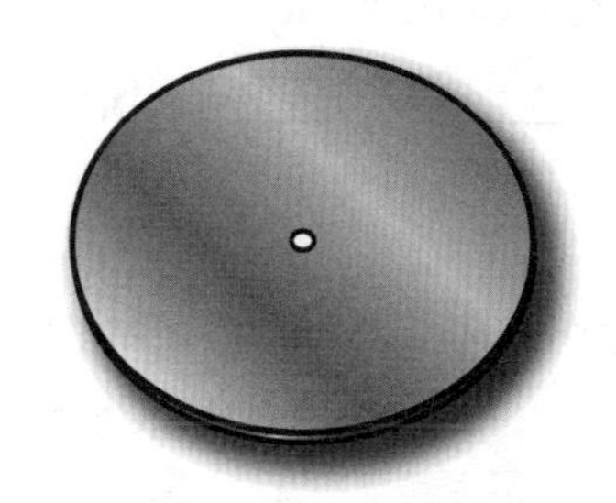

Step 3: Align ruler across the circle through the punched hole. Measure and score a line with the nail from each side of the circle toward the center. Stop scoring when you are 3⁄4″ away from center hole. Continue aligning ruler, measuring and scoring until you have eight equally placed score lines that are 3⁄4″ away from the center hole.

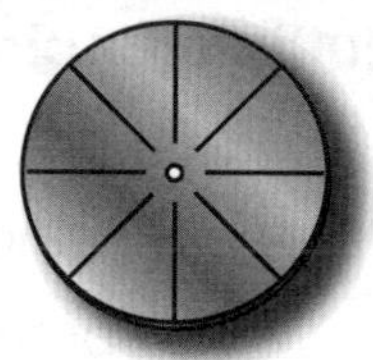

Step 4: Now, cut down the score lines. Slightly bend and twist each blade. Keep bending the blades in the same direction like a pinwheel.

Step 5: Push the pencil through the center hole. Cut a 20 inch piece of string and attach it to the pencil. Tie the nut to the other end of the string.

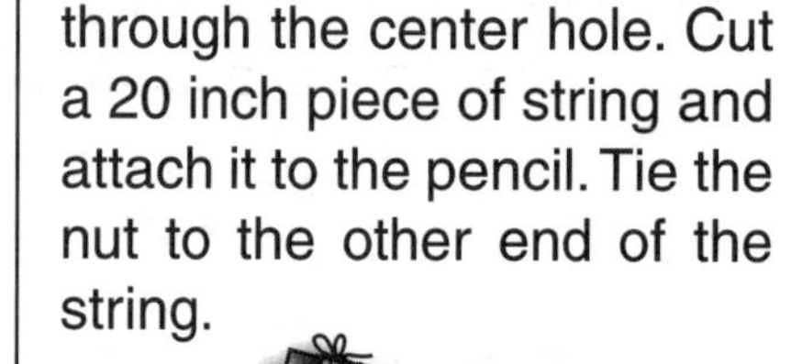

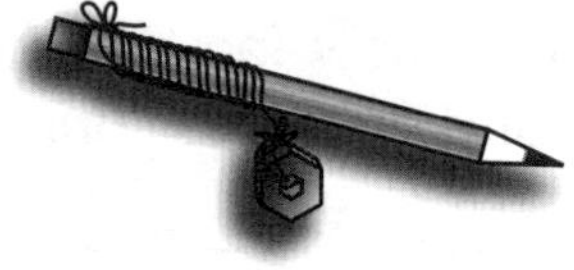

Step 6: Hold the pencil ends in your up-turned thumbs and forefin-gers. Place the blades under the running water.

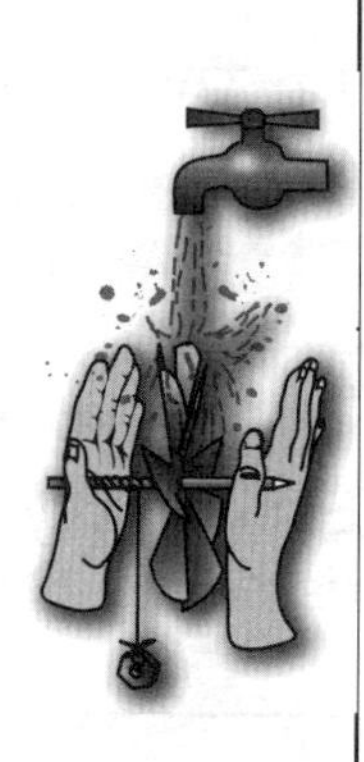

Observation: What happens to the nut? ______________________________

____

Conclusion: How is this process related to the production of hydroelectricity?

____

____

____

Name: ______________________________ Date: ______________________

Inquiry Lab: Hydro Power

Purpose: The purpose is a question that asks what you want to learn from the investigation. It should be clearly written, it usually starts with the verb "Does," and it can be answered by measuring something.

Purpose: *Does the height of a hole in a milk jug affect the distance of the water stream produced?*

Research: The goal of the research is to find information that will help you make a prediction about what will occur in your experiment. Investigate water pressure, dams, and hydroelectricity. Use the lines below for note taking.

____

____

____

____

____

____

____

____

____

____

____

____

Online Science: Learn more about hydroelectric energy at the following web-site. "Hydropower Basics—Energy from Moving Water." U.S. Department of Energy. <http://www.eia.doe.gov/kids/energy.cfm?page=hydropower_home-basics>

Hypothesis: Make an educated guess about what you think will happen in your project. Your hypothesis should be clearly written. It should answer the question stated in the purpose, be brief and to the point, and identify the independent and dependent variables.

Example: *The height of a hole in a milk jug* (choose one) ***will*** *or* ***will not*** *affect the distance of the water stream produced.*

Hypothesis: ____

____

Name: ______________________ Date: ______________

Procedure: The procedure is a plan for your experiment. The plan includes a list of the materials needed, step-by-step how-to directions (written like a recipe) for conducting the experiment, and it identifies the variables. Measurements are made and recorded using metric units.

Materials Needed:
half gallon plastic milk jug
water
10 p nail
duct tape
2 metric rulers
permanent marker
step stool

Variables
Independent: height of holes
Dependent: distance of stream
Constants: same amount of water, container, size of holes

Note: Do this activity outside, or cover the floor with newspaper to help absorb the water.

Experiment:
Controlled Setup:
Step 1: Measure 2 cm up from the bottom of the milk jug and mark with permanent marker.
Step 2: Using the 10 p nail, punch a hole in the center of the front of the jug at the 2-cm mark.
Step 3: Measure 4 cm up from the bottom of the jug, mark and punch a hole in the center above the 2-cm hole.
Step 4: Measure 8 cm up from the bottom of the jug, mark and punch a hole in the center above the 4-cm hole.
Step 5: Measure 16 cm up from the bottom, mark and punch a hole in the center above the 8-cm hole.
Step 6: Firmly tape each hole with a piece of duct tape.
Step 7: Place metric rulers end to end on the floor at the base of the step stool.
Step 8: Fill the jug with water and place on the edge of the step stool with the holes pointing toward the metric rulers.
Step 9: Hold the jug firmly and quickly remove the tape from the 2-cm marked hole. Record the distance at which the stream of water hits the floor in the data table. Repeat the test two more times.

Experimental Setup:
Repeat steps 6 through 9, measuring and recording the distance the water travels from the other three holes.

Results: Record the test results in the data table below in centimeters. Calculate the average distance for each hole height.

Controlled Setup				
Height of Hole	**Trial #1**	**Trial #2**	**Trial #3**	**Average**
2 cm				
Experimental Setup				
Height of Holes	**Trial #1**	**Trial #2**	**Trial #3**	**Average**
4 cm				
8 cm				
16 cm				

Name: ______________________ Date: ______________

Analysis: Study the results of your experiment. Create a graph that will compare the average distance traveled by the water in the control setup with the experimental setups. Place the dependent variable (distance of stream) on the *y*-axis. Place the independent variable (height of the holes) on the *x*-axis.

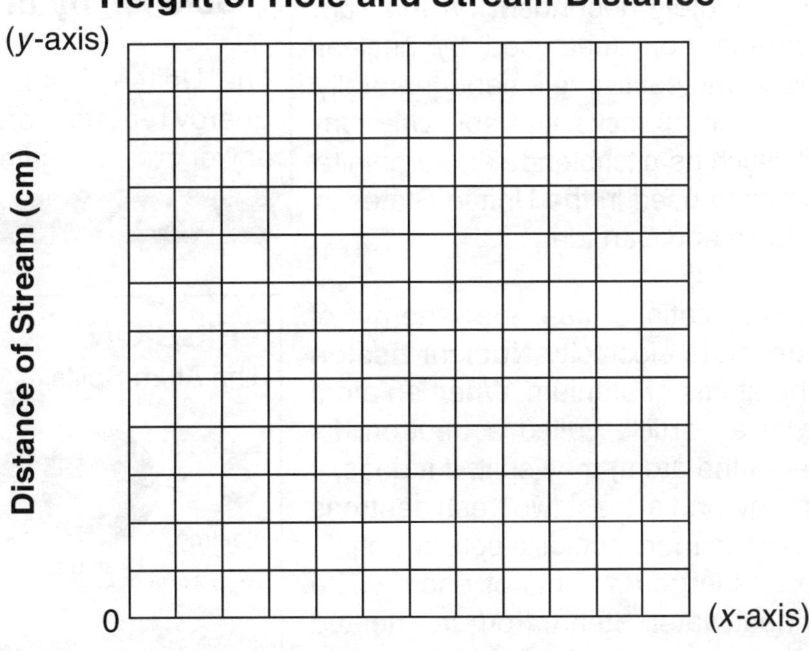

Height of Holes (cm)

Conclusion: Write a summary of the experiment (what actually happened). It should include the purpose, a brief description of the procedure, and whether or not the hypothesis was supported by the data collected. Use key facts from your research to help explain the results. The conclusion should be written in first person ("I").

Nuclear Energy
Student Information

Nuclear Fission

Nuclear power is the only major source of energy that is not ultimately dependent on the sun. **Uranium** is a nonrenewable fuel used by nuclear power plants. It is a radioactive element, naturally occurring in low levels in all rocks and soil. Uranium is found in minerals such as pitchblende and uraninite ore. Most of the uranium used in the United States is imported from Australia and Canada.

Science by the Numbers

The United States Department of Energy reports there are 65 nuclear power plants located in the United States.

Nuclear power stations use the energy of nuclear fission to generate electricity. **Nuclear fission** involves splitting the atoms of uranium. When an atom of uranium is hit by a particle called a neutron, its nucleus (the center of the atom) may split. If it does, it releases a lot of energy, and at least two fresh neutrons shoot off. They cause two more nuclei to split in a chain reaction. This releases large amounts of energy. The energy is used to heat water. Steam from the heated water turns turbines, which generate electricity.

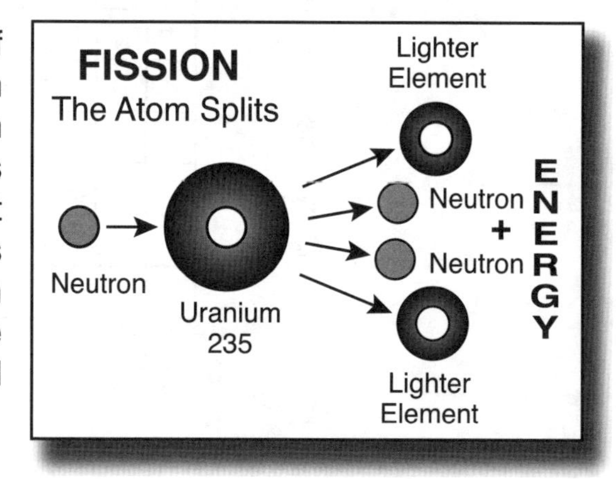

Early History of Nuclear Energy

- In 1789, Martin Klaproth, a German chemist, discovered that uranium was in the mineral called pitchblende.
- In the 1800s, the hazards of exposure to radiation were not yet known. People used uranium to make yellow glass. It was also used in paint and glaze for glass and ceramic objects.
- Uranium's radioactive properties were not noticed until 1896. Marie Curie named the new phenomenon **radioactivity**.
- In Italy in 1934, physicist Enrico Fermi demonstrated that neutrons could split many kinds of atoms.
- In 1942, the first self-sustaining nuclear chain reaction occured at the University of Chicago.
- In 1951, the first electric power from nuclear energy was produced in the Experimental Breeder Reactor I at a site in Idaho.
- In 1954, the USS *Nautilus*, the first nuclear-powered submarine, was launched.
- In 1955, Arco, Idaho, became the first town powered by a nuclear power plant.
- In 1986, the worst nuclear power plant disaster to date occurred at the Chernobyl nuclear power plant in the Ukraine, resulting in a total of 56 deaths directly, with more than 4,000 extra cancer deaths among those exposed.

Just the Facts

The first voyage under the North Pole icecap was made by a nuclear submarine.

The world's largest single uranium deposit is located at the Olympic Dam Mine in South Australia.

In 1979, America's worst nuclear power plant accident occurred in Pennsylvania at the Three Mile Island Nuclear Power Plant. Half the fuel melted in one of two nuclear reactors, releasing harmful radiation into the environment.

How Nuclear Energy is Used

Nuclear Power Plants: **Nuclear fission** takes place inside a **reactor** that is surrounded by a thick layer of concrete or steel. The center of the nuclear reactor is called the **core**. The core contains pipes called fuel rods and control rods. The **fuel rods** are filled with uranium. Nuclear fission takes place inside the fuel rods. The energy produced from the nuclear fission heats water in the core. The heated water is piped to a machine called a **heat exchanger**. The heat exchanger uses the hot water to boil a huge tank of water to produce steam. The steam is used to generate electricity. Round buildings called cooling towers outside the nuclear power plant release excess steam produced by the heat exchanger.

Nuclear Submarines: Some nuclear reactors are used to power submarines. The United States, Great Britain, China, and Russia are countries that have nuclear submarines.

Dangers of Nuclear Energy

The waste produced by nuclear power plants is **radioactive**. During the process of fission, particles harmful to humans, animals, and the environment are emitted as **radiation**. There is no way to turn **nuclear waste** into something harmless. Instead, it must break down on its own over thousands of years. The basic method of nuclear waste disposal is to place it in special containers. The containers are stored in nuclear repositories where they are then submerged in large pools of water to cool. Later, they are placed in reinforced casks or in concrete bunkers.

Future Use of Nuclear Energy

Scientists are hoping that the next generation of nuclear power stations will be powered by **nuclear fusion**. The fuel will be hydrogen, and the waste material will be water. The energy can be harnessed and used to generate electricity.

Advantages

- A nuclear power station produces no carbon dioxide.
- Nuclear energy can be man-made.

Disadvantages

- It is very expensive to build a nuclear power plant.
- Nuclear safety, long-term disposal of waste, and the danger of fuel being used to produce nuclear weapons are of concern to many people.

Name: ______________________________ Date: ______________

Check Point

Matching

_____ 1. uranium
_____ 2. nuclear fission
_____ 3. core
_____ 4. fuel rods
_____ 5. radiation

a. harmful particles emitted during fission
b. pipes in the core of a nuclear reactor filled with uranium
c. splitting the atoms of uranium
d. center of the nuclear reactor
e. fuel used by nuclear power plants

Fill in the Blank

6. Nuclear power is the only major source of energy that is not ultimately dependent on the ______________.
7. Nuclear power stations use the energy of nuclear fission to generate ______________.
8. Cooling towers outside the nuclear power plant release excess ______________ produced by the heat exchanger.
9. Nuclear safety, long-term disposal of waste, and the danger of fuel being used to produce ______________ ______________ are of concern to many people.
10. During the process of fission, particles harmful to animals and the environment are emitted as ______________.

Multiple Choice

11. Number of power plants located in the United States.
    a. 55 b. 77 c. 65 d. 99

12. The first United States nuclear-powered submarine.
    a. USS *Arizona* b. USS *Nautilus*
    c. USS *Missouri* d. USS *John F. Kennedy*

Constructed Response

Explain why using nuclear energy as an alternative energy source is a concern to many people. Give specific examples and details to support your answer.

____
____
____
____
____

Name: ______________________ Date: ______________

Mini Labs

Mini Lab #1: Chain Reaction

Materials:
marbles

Procedure: In a nuclear fission reaction in a nuclear power plant, the radioactive element uranium-235 is used in a chain reaction. Create a chain reaction. Set 30 marbles on a table. Place another marble on the table and flick it into the other marbles on the table.

Observation:

____

____

Conclusion:

How is the reaction of the marbles similar to a nuclear fission reaction in a power plant?

____

____

Mini Lab #2: Controlled Chain Reaction

Materials:
dominoes ruler

Procedure: The core of a nuclear reactor contains fuel rods and control rods. The control rods are used to slow down or stop the nuclear chain reaction. Create a controlled chain reaction. Arrange the dominoes in a line as shown at right. Now, knock over the first domino in the line. Observe the chain reaction. Set up the dominoes again. This time hold the ruler anywhere in the line between two of the dominoes. Knock over the first domino in the line.

Observation:

____

____

Conclusion:

How is the reaction of the dominoes similar to a nuclear fission reaction in a power plant?

____

____

Name: ______________________________ Date: ______________

Inquiry Lab: Radiation

Purpose: The purpose is a question that asks what you want to learn from the investigation. It should be clearly written, it usually starts with the verb "Does," and it can be answered by measuring something.

Purpose: *Does distance an object is from a radiation source affect the strength of the exposure?*

Research: The goal of the research is to find information that will help you make a prediction about what will occur in your experiment. Investigate nuclear energy, radiation, and Geiger counters. Use the lines below for note taking.

____

____

____

____

____

____

____

____

____

____

____

____

Online Science: Learn more about nuclear energy at the following website. "Energy Story: Chapter 13: Nuclear Energy—Fission and Fusion." California Energy Commission. <http://www.energyquest.ca.gov/story/chapter13.html>

Hypothesis: Make an educated guess about what you think will happen in your project. Your hypothesis should be clearly written. It should answer the question stated in the purpose, be brief and to the point, and identify the independent and dependent variables.

Example: *The distance an object is from a radiation source* (choose one) ***does*** *or* ***does not*** *affect the strength of the exposure.*

Hypothesis: ____

____

____

Name: ______________________________ Date: ____________________

Procedure: The procedure is a plan for your experiment. The plan includes a list of the materials needed, step-by-step how-to directions (written like a recipe) for conducting the experiment, and it identifies the variables. Measurements are made and recorded using metric units.

Materials Needed:

radiation source (low energy beta or gamma source)

(A Radioactive Source Kit can be ordered from the website below. This set of three radioactive sources is completely safe for students when used according to the directions, and includes one each of an alpha, beta, and gamma source. <http://sciencekit.com/radioactive-source-kit/p/IG0023988/>)

meter stick

Geiger counter (reading in counts/min or cpm)

Variables

Independent: distance from detector probe

Dependent: Geiger counter reading

Constants: Geiger counter, ruler, radiation source

Experiment:

Controlled Setup:

Step #1: Position the Geiger counter detector probe at one end of a meter stick.

Step #2: Place the radiation source 2 centimeters from the probe. Record the reading in the data table.

Experimental Setup:

Step #1: Reposition the radiation source at 2 centimeter increments farther from the probe and record the instrument reading at each distance. (Remember: keep the center of the source aligned with the center of the detector as the source is moved.)

Results: Record the distance (cm) and Geiger counter reading (counts/min) in the data table.

Controlled Setup	
Distance (cm)	**Reading (counts/min)**
Experimental Setup	
Distance (cm)	**Reading (counts/min)**

Name: ____________________ Date: ____________

Analysis: Study the results of your experiment. Plot the Geiger counter radiation readings versus distance on the graph. Place the dependent variable (radiation readings) on the *y*-axis. Place the independent variable (distance) on the *x*-axis.

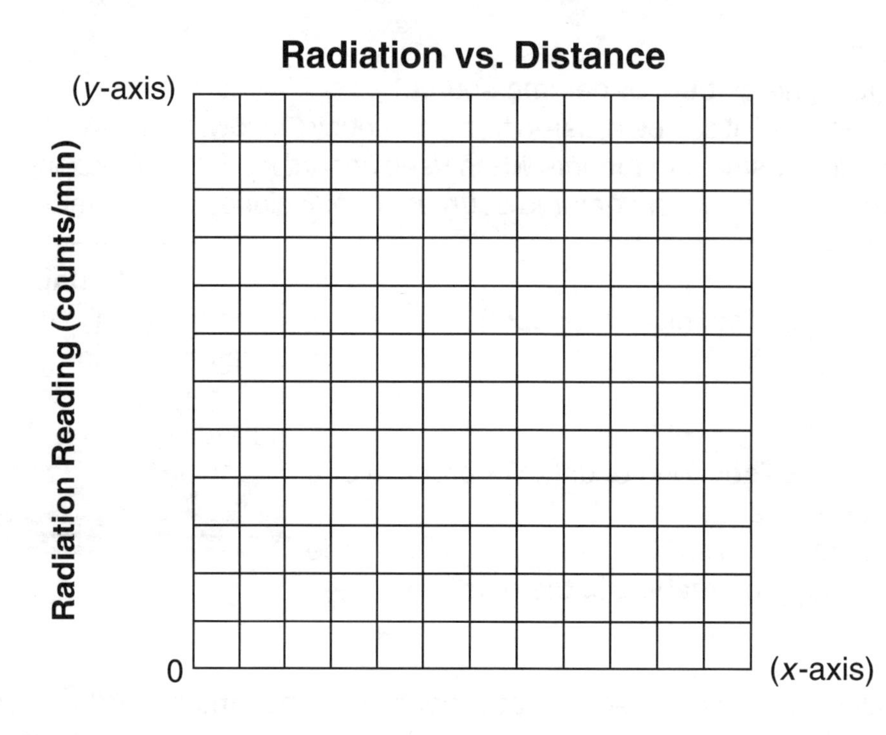

Conclusion: Write a summary of the experiment (what actually happened). It should include the purpose, a brief description of the procedure, and whether or not the hypothesis was supported by the data collected. Use key facts from your research to help explain the results. The conclusion should be written in first person ("I").

____

____

____

____

____

____

____

____

____

____

____

Answer Keys

Energy
Check Point (page 7)
Matching
1. e 2. c 3. b 4. a 5. d
Fill in the Blank
6. fission, fusion 7. wood 8. sun
9. Natural gas 10. fossil fuels
Multiple Choice
11. b 12. c
Constructed Response
A country's economic health depends on the availability of reliable and affordable energy. Vast amounts of energy are needed to support modern technological advancements. Energy is used to heat, cook, and light homes and businesses. Industries use energy to generate steam and heat for manufacturing. Gasoline and diesel are used for transportation. Agriculture uses diesel, gasoline, and propane.
Mini Labs (page 8)
Mini Lab #2 (Answers should be rounded to nearest hundredth.)
1. \$0.02 2. \$0.02 3. \$0.36 4. \$0.06
5. \$0.60 6. \$0.09 7. \$0.77
Inquiry Lab (pages 9–11)
Conclusion: Answers will vary but should include the following information: the purpose, a brief description of the procedure, and whether or not the hypothesis was supported by the data collected. (A brief shower uses less water.)

Energy Sources
Check Point (page 14)
Matching
1. b 2. e 3. a 4. c 5. d
Fill in the Blank
6. fossil fuels 7. electricity, liquid
8. exhaustible 9. Solar 10. Carbon dioxide
Multiple Choice
11. c 12. c
Constructed Response
There is a difference between inexhaustible resources and exhaustible resources. Inexhaustible resources cannot be used up by humans. Examples are solar, wind, ocean, and biomass. Exhaustible resources can be depleted in the future. Examples are fossil fuels.
Mini Labs (page 15)
Mini Lab #2, *Conclusion:* Oil spills have a negative effect on a marine ecosystem because they are difficult to contain, to clean up on shorelines, and to clean off of marine life.
Inquiry Lab (pages 16–18)
Conclusion: Answers will vary but should include the following information: the purpose, a brief description of the procedure, and whether or not the hypothesis was supported by the data collected. (Plants grown in neutral soil water will have longer roots.)

Electricity
Check Point (page 21)
Matching
1. c 2. e 3. b 4. a 5. d
Fill in the Blank
6. generator 7. protons, neutrons, electrons
8. electrical charge 9. Benjamin Harrison
10. kilowatt-hours
Multiple Choice
11. c 12. b
Constructed Response
Electricity is considered a secondary power source because it is generated from other energy sources. The energy sources we use to make electricity are called primary sources of energy. Coal, nuclear, and solar energy are examples of primary sources of energy.
Mini Labs (page 22)
Mini Lab #2
1. 5 hrs. 2. 1/2 or 0.5 hrs.
3. 2 1/2 or 2.5 hrs. 4. 4 hrs.
Inquiry Lab (pages 23–25)
Conclusion: Answers will vary but should include the following information: the purpose, a brief description of the procedure, and whether or not the hypothesis was supported by the data collected. (Incandescent bulbs produce the most heat, and LED bulbs produce the least heat.)

Wind Energy
Check Point (page 28)
Matching
1. b 2. a 3. e 4. d 5. c
Fill in the Blank
6. Wind, generator 7. Persia 8. electricity
9. greenhouse 10. energy
Multiple Choice
11. b 12. a
Constructed Response
Wind is considered an inexhaustible resource because it is unlimited, meaning that we will always have wind. Wind is air in motion; it's a result of the uneven heating of the earth's surface by the sun. Ancient people used it to power the sails of their boats. Today, wind turbines are used to generate electricity.

Mini Labs (page 29)

Mini Lab #1, Conclusion: It would be important to place the blades of a wind turbine facing the wind so the blades would turn.

Mini Lab #2, Conclusion: It would be important to place a wind turbine in an area that has a constant flow of high wind speeds so the turbines generate energy.

Inquiry Lab (pages 30–32)

Conclusion: Answers will vary but should include the following information: the purpose, a brief description of the procedure, and whether or not the hypothesis was supported by the data collected. (The number of blades affects voltage output. The two-blade model should produce the highest voltage.)

Solar Energy

Check Point (page 35)

Matching

1. b 2. e 3. a 4. d 5. c

Fill in the Blank

6. thermal, electrical 7. dish 8. Solar ponds
9. fossil-fueled 10. foreign

Multiple Choice

11. c 12. b

Constructed Response

Scientists predict the sun is capable of producing energy for billions of years. The sun is a big ball of gas that produces energy resulting from nuclear reactions in its core. In the core, hydrogen atoms combine to form helium atoms. This gives off radiant energy. The radiant energy travels to Earth as heat and light.

Mini Labs (page 36)

Mini Lab #1, Conclusion: Black paper and paint are used to construct the solar cooker because black absorbs light energy.

Mini Lab #2, Observations: 1. The balloon on the black bottle is inflated. 2. The black bottle feels warmer than the white bottle.

Conclusion: Dark colors absorb more light energy than lighter colors. The color black absorbs light energy. Inside the bottle heats up, causing the air molecules to start moving faster and further apart, expanding the balloon.

Inquiry Lab (pages 37–39)

Conclusion: Answers will vary but should include the following information: the purpose, a brief description of the procedure, and whether or not the hypothesis was supported by the data collected. (The color of the tubing affects the water temperature. Black tubing produced hotter water.)

Ocean Energy

Check Point (page 42)

Matching

1. d 2. c 3. b 4. e 5. a

Fill in the Blank

6. Tidal energy 7. tides
8. mechanical, electrical 9. inexhaustible source
10. Ocean Thermal Energy Conversion

Multiple Choice

11. d 12. a

Constructed Response

Since tides are caused by the gravitational pull between Earth and the moon, waves from the tides cannot be stopped. So tidal energy is an inexhaustible resource. Using this unlimited resource has led to the development of devices that capture this energy. Tidal barrages, tidal fences, and tidal turbines have been constructed to convert tide energy to electrical energy.

Mini Labs (page 43)

Mini Lab #1, Observation: Waves move through the water.

Conclusion: The energy gets stronger the faster the wave moves.

Mini Lab #2, Observation: The fan moves the water.

Conclusion: The stronger the wind, the bigger the waves.

Inquiry Lab (pages 44–46)

Conclusion: Answers will vary but should include the following information: the purpose, a brief description of the procedure, and whether or not the hypothesis was supported by the data collected. (The wave height decreases as the depth of water increases. The wave speed increases as the depth of the water increases.)

Biomass Energy

Check Point (page 49)

Matching

1. e 2. d 3. a 4. b 5. c

Fill in the Blank

6. Brazil 7. Biofuels 8. decaying
9. soybeans 10. fossil fuels

Multiple Choice

11. c 12. d

Constructed Response

Plants absorb carbon dioxide from the environment during photosynthesis. The fuels emit the same amount of carbon dioxide when burned as the amount absorbed by the plants during growth. So biofuels are considered carbon neutral.

Mini Labs (page 50)

Mini Lab #1, Observation: The balloon inflated.

Conclusion: The yeast feeds on the sugar in the plant material producing a gas and alcohol during the process of fermentation.

Mini Lab #2, Observation: The balloon inflated.

Conclusion: Decaying material in the landfill produces a gas that inflates the balloon.

Inquiry Lab (pages 51–53)

Conclusion: Answers will vary but should include the following information: the purpose, a brief description of the procedure, and whether or not the hypothesis was supported by the data collected. (Peanuts produce the most energy.)

Geothermal Energy

Check Point (page 56)

Matching

1. e 2. d 3. a 4. b 5. c

Fill in the Blank

6. heat pump 7. Binary cycle power plants
8. Italy 9. New Zealand 10. California

Multiple Choice

11. b 12. a

Constructed Response

Geothermal energy is considered an inexhaustible resource because heat is continually released from the core of the earth, and therefore, it is replaced naturally.

Mini Labs (page 57)

Mini Lab #1, Observation: The paper held over the candle has black streaks.

Conclusion: Steam energy is cleaner.

Mini Lab #2, Conclusion: The heat causes pressure to build up, allowing steam to escape from the bottle.

Inquiry Lab (pages 58–60)

Conclusion: Answers will vary but should include the following information: the purpose, a brief description of the procedure, and whether or not the hypothesis was supported by the data collected. (The pinwheel spins faster with more holes in the foil.)

Hydro Energy

Check Point (page 63)

Matching

1. c 2. a 3. e 4. b 5. d

Fill in the Blank

6. evaporation, condensation, precipitation
7. Norway 8. Hydroelectricity
9. mechanical 10. dams

Multiple Choice

11. b 12. b

Constructed Response

Research is focusing on building turbines that do not require damming because building new dams is too expensive, and wildlife and habitat are destroyed. These new turbines would allow fish to pass through them unharmed.

Mini Lab (page 64)

Mini Lab Observation: It is lifted or lowered by the string. Water power turns the turbine and pencil, making the string wrap around the pencil or unwrap.

Conclusion: It is similar to how water turns a turbine to generate electricity in a power plant.

Inquiry Lab (pages 65–67)

Conclusion: Answers will vary but should include the following information: the purpose, a brief description of the procedure, and whether or not the hypothesis was supported by the data collected. (The water stream distance is the greatest on the lowest hole.)

Nuclear Energy

Check Point (page 70)

Matching

1. e 2. c 3. d 4. b 5. a

Fill in the Blank

6. sun 7. electricity 8. steam
9. nuclear weapons 10. radiation

Multiple Choice

11. c 12. b

Constructed Response

Nuclear energy as an alternative energy source is a concern to many people. The waste produced by nuclear power plants is radioactive. Nuclear safety, long-term disposal of waste, and the danger of fuel being used to produce nuclear weapons are problems.

Mini Labs (page 71)

Mini Lab #1, Observation: The flicked marble hits a marble, causing that marble to move and hit other marbles.

Conclusion: Marbles hitting each other cause a chain reaction similar to nuclear fission. When an atom of uranium is hit by a neutron, its nucleus may split. Two fresh neutrons shoot off. They cause two more nuclei to split in a chain reaction.

Mini Lab #2, Observation: The rest of the dominoes are stopped from falling.

Conclusion: The control rods in the core stop a chain reaction similar to the way the ruler stopped the chain reaction with the dominoes.

Inquiry Lab (pages 72–74)

Conclusion: Answers will vary but should include the following information: the purpose, a brief description of the procedure, and whether or not the hypothesis was supported by the data collected. (The distance does affect the strength of the exposure. The farther away from the source of radiation, the less the exposure.)

Bibliography

"Biopower." United States Department of Energy. < http://www1.eere.energy.gov/biomass/>.

de Pinna, Simon. (2001) *Science Projects: Electricity.* London: Hodder Wayland.

"Energy Kids." Energy Information Administration. <http://tonto.eia.doe.gov/kids/>.

Freeman, S. David. (2007) *Winning Our Energy Independence.* Layton, UT: Gibbs Smith.

Gibson, Diane. (2004) *Sources of Energy: Geothermal Power.* North Mankato, MN: Smart Apple Media.

Hawkes, Nigel. (2003) *Saving Our World: New Energy Sources.* London: Franklin Watts.

"Hydropower." United States Department of Energy. <http://tonto.eia.doe.gov/kids/energy.cfm?page=hydropower_home-basics>.

"Make a Windsock." Weather Wiz Kids. (2009) <http://weatherwizkids.com/experiments-windsock.htm>.

Margulies, Phillip. (2006) *This is Your Government: The Department of Energy.* New York: The Rosen Publishing Group, Inc.

Morgan, Sally. (2009) *Science at the Edge: Alternative Energy Sources.* Chicago: Heinemann Library.

"Ocean Energy." Minerals Management Service, U.S. Department of the Interior. <http://www.mms.gov/mmsKids/PDFs/OceanEnergyMMS.pdf>.

Richards, Julie. (2003) *Future Energy: Solar Power.* North Mankato, MN: Smart Apple Media.

Richards, Julie. (2004) *Future Energy: Water Power.* North Mankato, MN: Smart Apple Media.

"Science Projects: Geothermal Power Plant Model." California Energy Commission. <http://www.energyquest.ca.gov/projects/geothermal-pp.html>.

"Science Projects: The Force of Water." California Energy Commission. <http://www.energyquest.ca.gov/projects/hydro-power.html>.

ScienceSaurus: A Student Handbook. (2002) Wilmington, MA: Great Source Education Group.

"Solar Cooking." Solar Cookers International. <http://solarcooking.org/plans/plans.pdf>.

"The History of Nuclear Energy." United States Department of Energy. <http://www.ne.doe.gov/pdfFiles/History.pdf>.

"The Infinite Power of Texas: Generation for Generations." Texas State Energy Conservation Office. <http://www.infinitepower.org/>.

"The Power of Water." Niagara Pennsylvania Conservation Authority. <http://www.niagarachildrenswaterfestival.com/images/pdf/Water Attitude Teacher Resources.pdf>.

Taylor, Barbara. (1993) *Energy and Power.* New York: Franklin Watts.

"Wave Size and Depth." Southeast Coastal Ocean Observing Regional Association. <http://secoora.org/classroom/waves/activities/wave-size-and-depth/>.

"Wind Machine Instructions." Public Broadcasting Service. <http://www.pbs.org/now/classroom/windmachine.pdf>.

Zoelliner, Tom. (2009) *Uranium: War, Energy, and the Rock that Shaped the World.* New York: Viking.